TECHNIQUES FOR THE STUDY OF MIXED POPULATIONS

THE SOCIETY FOR APPLIED BACTERIOLOGY
TECHNICAL SERIES NO. 11

TECHNIQUES FOR THE STUDY OF MIXED POPULATIONS

Edited by

D. W. LOVELOCK
H. J. Heinz Co. Ltd, Hayes Park, Hayes, Middlesex, England

AND

R. DAVIES
National College of Food Technology, Weybridge, Surrey, England

1978

ACADEMIC PRESS
LONDON · NEW YORK · SAN FRANCISCO
A Subsidiary of Harcourt Brace Jovanovich, Publishers

ACADEMIC PRESS INC. (LONDON) LTD
24/28 OVAL ROAD
LONDON NW1

U.S. Edition Published by
ACADEMIC PRESS INC.
111 FIFTH AVENUE
NEW YORK, NEW YORK 10003

Library of Congress Catalog Card Number: 78–67895
ISBN: 0–12–456650–2

Printed in Great Britain by
Latimer Trend & Company Ltd, Plymouth

Contributors

B. W. ADAMS, *ARC Food Research Institute, Colney Lane, Norwich NR4 7UA, Norfolk, England*

P. A. AYRES, *Ministry of Agriculture, Fisheries and Food, Fisheries Laboratory, Burnham-on-Crouch, Essex, England*

ELLA M. BARNES, OBE, *ARC Food Research Institute, Colney Lane, Norwich NR4 7UA, Norfolk, England*

C. M. BROWN, *University of Dundee, Dundee DD1 4HN, Scotland*

H. W. BURTON, *Ministry of Agriculture, Fisheries and Food, Fisheries Laboratory, Burnham-on-Crouch, Essex, England*

G. S. COLEMAN, *ARC Institute of Animal Physiology, Babraham, Cambridge CB2 4AT, Cambridgeshire, England*

MARY L. CULLUM, *Ministry of Agriculture, Fisheries and Food, Fisheries Laboratory, Burnham-on-Crouch, Essex, England*

C. R. CURDS, *British Museum (Natural History), Cromwell Road, London, England*

D. C. ELLWOOD, *Microbiological Research Establishment, Porton, Wiltshire, England*

J. C. FRY, *University of Wales, Institute of Science and Technology, King Edward VII Avenue, Cardiff, CF1 3NU, Wales*

P. N. HOBSON, *The Rowett Research Institute, Bucksburn, Aberdeen AB2 9SB, Scotland*

R. S. HOLDOM, *University of Strathclyde, George Street, Glasgow G1 1XW, Scotland*

R. W. HORSLEY, *Freshwater Biological Association, The Ferry House, Ambleside LA22 0LP, Cumbria, England*

J. R. HUNTER, *Microbiological Research Establishment, Porton, Wiltshire, England*

N. C. B. HUMPHREY, *University of Wales Institute of Science and Technology, King Edward VII Avenue, Cardiff CF1 3NU, Wales*

C. S. IMPEY, *ARC Food Research Institute, Colney Lane, Norwich NR4 7UA, Norfolk, England*

V. F. LARSEN, *University of Strathclyde, George Street, Glasgow G1 1XW, Scotland*

M. J. LATHAM, *National Institute for Research in Dairying, Shinfield, Reading RG2 9AT, Berkshire, England*

T. LOVEGROVE, *International Marine Coatings, Biological Laboratory, Yealm Road, Newton Ferrers, Plymouth PL8 1BN, Devon, England*

G. C. MEAD, *ARC Food Research Institute, Colney Lane, Norwich NR4 7UA, Norfolk, England*

R. W. A. PARK, *University of Reading, London Road, Reading RG1 5AQ, Berkshire, England*

D. MC.L. ROBERTS, *British Museum (Natural History) Cromwell Road, London, England*

M. J. SPIVEY, *University of Strathclyde, George Street, Glasgow G1 1XW, Scotland*

R. SUMMERS, *The Rowett Research Institute, Bucksburn, Aberdeen AB2 9SB, Scotland*

M. TODD, *University of Strathclyde, George Street, Glasgow G1 1XW, Scotland*

L. G. WILLOUGHBY, *Freshwater Biological Association, The Ferry House, Ambleside LA22 0LP, Cumbria, England*

M. J. WOLIN, *Environmental Health Centre, New York State Department of Health, Albany, New York, NY USA*

CHIH-HUA WU, *British Museum (Natural History), Cromwell Road, London, England*

Preface

THIS volume includes contributions to the Autumn Demonstration Meeting of the Society for Applied Bacteriology held in October 1975 at the National College of Food Technology, St. Georges Avenue, Weybridge, Surrey. It is Number 11 in the Technical Series and continues the Society's policy of encouraging members and guests to exhibit methods and techniques which they have found to be most useful in the day-to-day work of their laboratories. The demonstrators have described their methods in this book which is intended to be a reference source at the bench. We wish to thank them all for the great care which they took both in the preparation of their demonstrations and for their contributions in this book.

Our thanks go also to the governing body and the academic and technical staff of the National College of Food Technology for permission to hold the meeting in their laboratories and for all their help with the arrangements.

D. W. LOVELOCK
R. DAVIES

October 1978

Contents

Techniques for the Study of Bacteria Epiphytic on Aquatic Macrophytes

J. C. Fry and N. C. B. Humphrey

University of Wales Institute of Science and Technology, Cardiff, Wales

Introduction

Aquatic macrophytes are found in freshwater, estuarine and marine habitats and can be submersed, floating or emergent. Emergent macrophytes have most of their leafy parts above water and are merely rooted in aquatic habitats, being found most often as marginal vegetation. The techniques described in this chapter deal mainly with the study of mixed populations of bacteria which are epiphytic on submersed and floating aquatic macrophytes found in freshwaters. Similar methods can probably be used for the study of the macrophytes found in estuarine or marine habitats.

There are many species of submersed and floating aquatic plants (Sculthorpe, 1967). Some are completely submersed, others have submersed and floating leaves and the rest have only floating parts. Some species have few roots but others have extensive rooting or rhizome systems which may or may not be anchored in the sediment. Many species have submersed leaves that are finely divided but some species have undivided submersed leaves whilst most floating leaves are also relatively undivided. However, all these aquatic macrophytes provide a large surface area for the growth of epiphytic bacteria.

It is possible that bacteria epiphytic on aquatic plants play an important role in aquatic ecosystems; although few studies of them have been made there is evidence of a large and active population (Allen, 1971; Ramsay and Fry, 1976). It has been shown by several groups of workers that planktonic algae secrete a large part of their photosynthetic product (Berman and Holm-Hansen, 1974). However, submersed aquatic macrophytes are sometimes more productive than are phytoplankton (Wetzel, 1975) and a significant part of their photosynthetically fixed carbon can also be secreted (Wetzel, 1969; Wetzel and Allen, 1972; Hough and

Wetzel, 1975). It is well known that the roots of terrestrial plants exude organic material and it is probable that aquatic plant roots do the same. The epiphytic bacteria are thus ideally placed to utilize these dissolved organic materials before they reach the water column. Work in this laboratory has shown that at times in a small pond there are more bacteria on the macrophytes than in the water and that these epiphytic bacteria are potentially capable of mineralizing more glucose than are the planktonic bacteria. Thus from the limited evidence available it seems likely that epiphytic bacteria are important in at least some freshwater ecosystems.

Sampling

Although weed-cutting machines have been used for sampling macrophytes and their associated invertebrate fauna (Gillespie and Brown, 1966) their use is unnecessary when studying bacterial epiphytes. Mixed samples may be obtained with simple grabs on handlines when sampling from the bank, or hand-cut samples can be obtained when a boat is available. With care hand-cut samples may be obtained from more precise locations and at known depths in water up to about 2 m. Samples near the surface can be taken with short or long handled shears and deeper samples obtained with the use of a face mask and snorkel tube, although a wet suit may be necessary in cold water. Samples from water up to 8 m in depth have been obtained with an Ekman dredge, various grapples and a specially designed quantitative macrophyte sampler (Rich *et al.*, 1971).

After plant material has been obtained it should be protected from dessication during transport to the laboratory in either plastic bags or wide-necked bottles with close-fitting lids. Well-cleaned containers and shears can be used, as sterile apparatus is unnecessary due to the large numbers of bacteria present on the plants. All subsequent techniques should be carried out immediately after the return of the samples to the laboratory. Although it is possible to wash the collected macrophytes free of loosely associated bacteria with sterile water before subsequent treatment, we believe it is preferable to use untreated plants so that the bacterial epiphytes remain undisturbed.

Although whole shoots and growing tips of macrophytes have been used for studies of macrophyte productivity (Adams *et al.*, 1974) studies of bacterial epiphytes have more commonly involved the use of plant fragments. Removed leaves of a variety of ages have been used in studies of the epiphytes of *Elodea canadensis* (Ramsay, 1974; Ramsay and Fry, 1976) and various stem and leaf parts have been used in studies of other plants (Allen, 1971; Ramsay and Fry, 1976). Allen (1971) has used discs of plant tissue removed with a cork borer in some of his studies. Although such plant parts are suitable for comparisons between epiphytes on different plants and for studies of epiphytes on largely mono-specific stands of aquatic macrophytes, we favour the use of mixed samples of plant parts if results are to be related to freshwaters with mixed stands of macrophytes.

Expression of Results

Results describing either the numbers or activity of bacterial epiphytes on aquatic plants are best expressed as dry weight, fresh weight or surface area of macrophyte. Some workers have expressed numbers as a function of volume of either homogenized macrophyte (Chan and McManus, 1967) or water squeezed from the plant (Coler and Gunner, 1969). Although this is satisfactory for comparative work within a single study, it makes comparisons between different workers and studies very difficult. Measurement of dry weights, fresh weights and surface areas of aquatic plants is not always straightforward, and as such problems are not often encountered by microbiologists a discussion of suitable methods is worthwhile.

Dry weight measurement

The plants are subsampled before microbiological examination and the dry weight of a known fresh weight of plant material is obtained by heating at 105° until constant weight is obtained. If the fresh weight of plant used in the microbiological procedures is known the dry weight can be found by proportion. Providing both sample and subsample of aquatic plant have the same amount of surface water no large errors will be involved.

Fresh weight measurement

Estimation of the true fresh weight of aquatic plants is more difficult as they are covered with a water film and retain a variable amount of water between their leaves and stems. Koegel *et al.* (1972) have shown that for *Myriophyllum spicatum* this surface water can be as much as 45% of the fresh weight. This water has to be removed before a true fresh weight can be obtained and since squeezing and draining will remove a variable amount of surface water they cannot be recommended. For small samples blotting with sterile adsorbant paper proves satisfactory but for larger samples spinning for one minute in a domestic spin-dryer at about 130 *g* has been successfully used to remove this water (Edwards and Owens, 1960; Westlake, 1968).

Surface area measurement

Direct methods

The surface area of most aquatic macrophytes with undivided leaves can be measured directly by first tracing the leaf outline on to graph paper and then calculating its surface area. With small undivided leaves a similar method can be used after the leaf outline has been enlarged on to the paper with an overhead projector. The surface area of such an outline is estimated by either counting squares on the graph paper, by weighing the cut out shape of the leaf or with a planimeter. For divided leaves the problem is more difficult but most leaves can be divided up into straight segments, the average length and width of each segment is measured microscopically and the surface area calculated. The surface areas of stems are easily measured with a similar technique. These methods have been used successfully for a variety of aquatic plants (Rosine, 1955; Harrod and Hall, 1962) but are difficult to perform aseptically and consequently such measurements are best performed on a subsample of known fresh weight. Ramsay and Fry (1976) have found that for leaves of *E. canadensis* the sum of the length and width is proportional to the surface area and consequently these parameters were measured for each leaf used in the study. The surface area of the leaves was determined with reference to standard regression lines obtained at intervals in the study. This involves a minimum of handling and consequently is easier to perform aseptically.

Indirect methods

In one study of the epiphytes of aquatic plants an indirect method of measuring surface area has been used (Allen, 1971). The method devised by Harrod and Hall (1962) involves immersion of the leaf in acetone, drying it completely and weighing the dry leaf. The leaf is then immersed in a standard solution of Teepol in 500 ml of water (Harrod and Hall, 1962) or a solution of Teepol 610 and a 50% dilution of Tween 80 (Allen, 1971) and is then removed and shaken for 20 s before it is re-weighed. This allows the weight of an even film spread over the leaf surface to be obtained. The surface area of each leaf is then found with reference to standard regression lines relating weight of surface film to leaf area. These regression lines must be obtained for each plant species studied using a direct method to assess surface area.

As fresh weight and surface area are comparatively difficult to measure, and workers express results in different ways, it would be convenient if dry weight, fresh weight and surface area were related in a constant manner for all species of aquatic plants. Results obtained in this laboratory (Table 1) show this not to be the case for the six species of submersed aquatic plants examined. Of the three ratios obtained that of

TABLE 1. Fresh weight, dry weight and surface area ratios for six species of aquatic macrophyte

Aquatic plant	Ratios ± 95% confidence intervals: Dry weight / fresh weight ($g\ g^{-1}$)	Surface area / fresh weight ($cm^2\ g^{-1}$)	Surface area / dry weight ($cm^2\ g^{-1}$)
Myriophyllum spicatum	0·098 ± 0·010	63·6 ± 6·7	653 ± 71
Apium inundatum	0·067 ± 0·015	99·7 ± 3·8	1485 ± 106
Apium nodiflorum	0·052 ± 0·006	85·1 ± 26·6	1616 ± 365
Potamogeton natans	0·145 ± 0·021	99·1 ± 52·4	685 ± 355
Elodea canadensis	0·156 ± 0·029	84·2 ± 2·5	548 ± 110
Callitriche sp.	0·062 ± 0·010	83·8 ± 4·1	1378 ± 299
Mean for all plants	0·096 ± 0·019	85·7 ± 6·7	1078 ± 215

Ratios are means of 3 or 4 values; 95% confidence intervals were calculated from untransformed data. Surface areas were measured by an appropriate direct method and fresh weights were obtained after blotting.

surface area to fresh weight was most constant, although some significant differences ($p = 0{\cdot}05$) were evident. The other two ratios showed many significant differences between species and in both cases *Apium inundatum*, *A. nodiflorum* and *Callitriche* sp. were all significantly different from each of the other plants. However, the mean value of the surface area to dry weight ratio (1078 $cm^2\ g^{-1}$, 0·93 mg cm^{-2}) agrees well with the value of 1 mg cm^{-2} obtained by Edwards and Owens (1965).

Results in which numbers of viable heterotrophs have been expressed as functions of dry weight, fresh weight and surface area (Table 2) show that the method of expression can result in different conclusions being drawn from the counts. The plants can be placed in a different order for each method of expression when ranked according to the size of their bacterial epiphyte populations. Tukey's test for significant differences ($p = 0{\cdot}05$) between means was used to examine the results more closely and showed many similarities in these data. Some groups of plants were not significantly different when their epiphyte populations were compared with any method of expression. These were

(1) *A. nodiflorum*, *E. canadensis* and *Callitriche* sp.;

(2) *A. inundatum* and *A. nodiflorum*;
(3) *M. spicatum*, *A. inundatum* and *Potamogeton natans*.

Also with all methods of expression *M. spicatum* and *A. inundatum* had significantly more epiphytes than did *Callitriche* sp. There were also differences in the data; with the dry weight results *M. spicatum* and *A. inundatum* had significantly more epiphytes than did *E. canadensis* but these differences were not found with the other two methods of expression. Conversely only when the results were expressed by surface area or fresh weight were there significant differences between epiphyte populations on *Callitriche* sp. and *P. natans* and between these populations on *A. nodiflorum* and both *P. natans* and *M. spicatum*.

TABLE 2. Numbers of viable heterotrophic bacteria on six species of aquatic macrophytes expressed in three different ways

Aquatic plant	Number of viable heterotrophs ± 95% confidence intervals		
	g^{-1} fresh weight of plant material × 10^7	g^{-1} dry weight of plant material × 10^8	cm^{-2} surface area of plant material × 10^5
Myriophyllum spicatum	3·42 ± 2·05	3·49 ± 2·10	5·38 ± 3·24
Apium inundatum	2·38 ± 0·52	3·53 ± 0·77	2·38 ± 0·52
Apium nodiflorum	1·13 ± 0·83	2·15 ± 1·78	1·33 ± 1·10
Potamogeton natans	4·36 ± 2·79	2·99 ± 1·94	4·41 ± 2·82
Elodea canadensis	1·92 ± 1·08	1·24 ± 0·69	2·28 ± 1·28
Callitriche sp.	0·84 ± 0·36	1·34 ± 0·62	1·01 ± 0·44

Numbers of viable heterotrophs are means of 4 or 5 values obtained in January and February 1975 after removal from the plants by stomaching and counted on CPS agar incubated at 18° for eight days; 95% confidence intervals were calculated from untransformed data. Surface areas were obtained by an appropriate direct method and fresh weights were obtained after blotting.

The overall conclusion from these statistical comparisons is that the pattern of significant differences is identical when the epiphyte counts were expressed by fresh weight or surface area, but results expressed as dry weight show a different pattern. This demonstrates that care must be taken when drawing conclusions from results expressed in different ways.

It is not clear which method of expressing the numbers and activity of bacterial epiphytes is best, but probably the method of choice will depend on the study in hand. If the aim of the study is to compare epiphytic, planktonic and benthic communities then results are probably best expressed by fresh or dry weight, as biomass estimates of macrophytes are normally expressed in a similar manner (Sculthorpe, 1967; Wetzel, 1975). However, as fresh weights are difficult to obtain and are subject to greater variability, results expressed as dry weight might well be preferable. Conversely, if plants are considered as surfaces for bacterial growth and comparisons of epiphyte populations on different plants are undertaken, it might be best to express results on a surface area basis.

Direct Observation

Direct observation of bacteria on the surface of aquatic plants have rarely been reported in the literature for the quantitative estimation of bacterial epiphyte populations. However, such techniques can give some useful qualitative information and bacteria can be observed directly by any of the following techniques.

Unstained material can be observed by phase contrast microscopy in simple wet mounts. This method of observation works well for macrophytic algae and for aquatic angiosperms with finely divided leaves. However, it is of little use for the observation of epiphytes on plants with large or thickened stems or leaves.

The epiphytes on the surface of any aquatic plant can also be observed by incident light fluorescence microscopy of leaves stained with acridine orange. Whole or parts of leaves are stained in acridine orange (120 μg ml^{-1}) for 3–5 min. The leaves are then removed and, after washing in distilled water, are either mounted in water under a coverslip or blotted dry, covered in immersion oil and viewed directly on a microscope slide. The use of incident light microscopy allows the observation of epiphytes on any thickness of leaf or other plant part. The authors have found the use of a Zeiss Universal Microscope with HBO 200 or HBO 50 mercury pressure-lamp, two BG12 exciter filters and barrier filter No. 50 to be a satisfactory microscope system, but others which have been used for direct counts of water bacteria on membrane filters (Jones, 1974; Jones and Simon, 1975) should prove satisfactory. Although qualitative observation is relatively straightforward with this technique, enumeration would be very difficult due to fading of the fluorescence during illumination and the depth of field problems encountered with high resolution lenses which are accentuated by the unevenness of the leaf surface.

Epiphytic algae and bacteria have been observed with scanning and transmission electron microscopy (Allanson, 1973). However, although the results show the algae clearly the bacterial epiphytes are harder to see and these techniques do not seem to be useful at present.

Another technique is currently being developed at the Freshwater Biological Association (Hŏssell and Baker, 1976). The method involves the staining of whole or parts of leaves with Phenolic Alanine Blue (Jones and Mollison, 1948) and direct observation of the epiphytes by bright field microscopy. This technique shows promise and is being used to count epiphytes on plant surfaces directly.

Removal of Bacteria

It is easier to study bacteria when they are in suspension than when they are attached to surfaces and consequently it is often preferable to remove the epiphytes from the plant before subsequent examination. This has been done by a variety of techniques which include squeezing (Coler and Gunner, 1969), washing (Edwards and Owens, 1965), swabbing (Potter, 1964; Allen, 1971), scraping (Ramsay, 1974), homogenizing (Chan and McManus, 1967; Laycock, 1974) and stomaching (Ramsay and Fry, 1976). For quantitative studies homogenizing and stomaching appear to be the best methods of epiphyte removal and these techniques will be described in more detail.

Homogenizing

Chan and McManus (1967) have critically examined the use of homogenization for the removal of bacterial epiphytes from macrophytic marine algae. Their conclusions probably also apply to the removal of epiphytes from freshwater macrophytes. It is first necessary to cut up the material into small pieces to avoid wrapping of stems around the homogenizer blade, since if this occurs epiphyte removal is inefficient. Although Chan and McManus (1967) do not report the size of the plant material used, Laycock (1974) using a similar technique used twenty 16 cm^2 pieces of *Laminaria longicruris* frond in a 4 oz blender jar with 100 ml of sterile water. There are two factors which reduce bacterial epiphyte numbers during homogenization. These are increase in temperature, above 30° for marine bacteria, and mechanical injury due to excessive shear forces during homogenization at speeds over 6000 r/min. Consequently, they recommend the use of a blender with a motor and blade assembly mounted above the blender jar. This apparatus does not allow the temperature to rise above 30° until between five and ten minutes of homogenizing. When a bottom mounted motor was used the temperature of the homogenate rose to 30° within one minute which was unsatisfactory. From their results, homogenizing for five minutes at 5000–6000 r/min was recommended. In Laycock's study these conclusions were confirmed by her preliminary experiments and she homogenized for five minutes at 3000 r/min.

Stomaching

Stomaching was recently devised for preparing suspensions of food and is as effective as homogenizing for the recovery of viable bacteria (Sharpe and Jackson, 1972; Sharpe, 1973). The sample is placed with diluent in a sterile plastic bag, which is put inside the stomacher and the bag is vigorously pounded on its outer surface by paddles inside the machine. This action removes even deep-seated bacteria with a negligible temperature rise. Work in this laboratory (Ramsay and Fry, unpublished) has compared the efficiency of the stomacher and homogenizer for the removal of epiphytic bacteria. Ten such comparisons have shown that in all but two cases stomaching gave higher viable counts than did homogenizing and that in three of these comparisons the increases were significant. This work has also shown that stomaching aquatic plants for five minutes is preferable to the 30 s used for food, as the increased time also increased the viable count in the epiphyte suspension.

Two stomaching procedures have been used in this laboratory for the removal of epiphytic bacteria. In the first procedure 60 g fresh weight of mixed aquatic plants are put into a sterile plastic bag (18 × 31 cm; A. J. Seward and Co. Ltd, 6, Stamford Street, London SE1 9UG) with 300 ml of sterile distilled water and this is then treated in a Colworth Stomacher-400 (A. J. Seward and Co. Ltd) for five minutes. The resulting epiphyte suspension is separated from the plant tissue by filtration through 380 μm bolting silk (John Staniar and Co., Sherborne Street, Manchester M3 1FD). The tissue is washed with a further 300 ml of sterile distilled water and the dry weight of the plant material determined.

If results are to be expressed on a surface area basis it is preferable to use smaller quantities of plant material. In these cases an alternative procedure is used in which 12–60 leaves of *E. canadensis*, 12 segments of *Chara vulgaris* or 4 leaves of *P. natans* are treated for five minutes with 50 ml of sterile water in a plastic bag (18 × 31 cm). In these cases, however, it was found necessary to put an additional buffer bag containing 200 ml of water but no plant material in the stomacher during treatment. This procedure gives higher viable counts of bacterial epiphytes than other stomaching procedures using small quantities of plant material and is consequently to be preferred.

It is probable that stomaching is a better procedure for the removal of bacterial epiphytes than homogenizing because of the higher viable counts obtained. Stomaching also has many practical advantages as it is quick and easy to use aseptically, little heat is generated, the plant tissue need not be cut up before treatment and remains fairly intact, giving an extract with little suspended plant debris.

Direct Microscopic Enumeration of Removed Epiphytes

Most microbiologists accept that from both theoretical and practical considerations viable counting procedures always underestimate the true population size. The development of adequate direct counting techniques was a response to this acknowledgement, but seems to have been little used to study removed epiphytes.

Available direct counting methods are of two general types.

Non-fluorochrome staining methods

Ramsay (1974) describes a method she used to count bacteria removed by homogenization from *E. canadensis*. The sample is filtered on to a Millipore membrane, stained with aqueous methylene blue (0·3%), and the membrane cleared with dimethyl sulphoxide in isopropanol. Comparing her counts with viable heterotrophs grown on nutrient agar (Oxoid), she found the viable numbers to be 1% of the total count on moribund *E. canadensis*; 3·8% of the total count for mature parts of the plant and 25·8% of the total count on the young apical leaves.

Fluorochrome staining methods

Fluorochrome methods, especially those using acridine orange-based dyes, have been widely used to enumerate bacteria in a range of habitats. Sorokin and Overbeck (1972) describe a method where acridine orange at a dilution of 1:20 000 to 1:50 000 is mixed with the sample in a counting chamber. When examined with transmitted light with a wavelength of 450 nm, bacteria should be seen as bright green fluorescent objects. Fenchel (1970) used a similar method to count total bacterial numbers on dead leaves of *Thalassia testuclinum* occurring as detritus. He found a range of counts between 1 and 9×10^6 cm^{-2} of surface area. The average bacterial number was approximately 3×10^6 cm^{-2} which, assuming a conversion factor of 1 $cm^{-2} \equiv 1$ mg^{-1} dry weight (Edwards and Owens, 1965), is equivalent to 3×10^9 g dry weight of detritus. This corresponds closely to the numbers of bacteria found by Ramsay and Fry (1976) on leaves of mature *E. canadensis*, using an incident light epifluorescence technique based on that of Jones (1974). This type of method has been favoured for obtaining total bacterial counts in water.

We have made use of a similar technique to enumerate bacteria directly from stomacher derived epiphyte suspensions. Twenty-five millilitres of an appropriate dilution of removed epiphytes in sterile distilled water is mixed for three minutes with 1 ml of acridine orange (120 mg litre^{-1}). An aliquot (0·5 to 25 ml) of stained suspension is then filtered through a black membrane filter of 0·45 μm mean pore size (Sartorius cat. No. SM 13006), under vacuum. When the aliquot size is less than 5 ml, 10 ml of sterile distilled water is dispensed on the membrane before addition of the sample, to ensure even distribution of bacteria on the membrane during filtration. The filter is then washed with an equal volume of sterile distilled water to remove excess stain. The membrane is supported on a microscope slide and examined with a Zeiss planapochromat × 63 (NA = 1·40) oil immersion objective, in a Zeiss Universal microscope. The membrane is illuminated with incident light from an HBO × 200 mercury pressure lamp, with two BG12 exciter filters and a No. 50 barrier filter. Green fluorescing, bacterial-shape objects in the central four squares of a 10 × 10 eye-piece grid are counted, in each of five fields in such a way that the whole area of the membrane is sampled. Two membranes are counted for each sample and assuming a Poisson distribution (Jones, 1974) the 95% confidence intervals calculated according to the expression

$$\bar{X} \pm t_n \sqrt{\frac{\bar{X}}{n}}$$

(Elliot, 1971).

Daley and Hobbie (1975) and Jones and Simon (1975) have examined critically the currently available epifluorescence techniques. Although they counted bacteria in water their findings were similar to those found in this laboratory for epiphyte suspensions, and a summary of all these results is worthwhile.

(1) Acridine-based fluorochromes used at a final concentration of 5 to 10 mg litre^{-1} for 3 to 5 min, give optimal results. The best final concentration for epiphyte suspensions is 5 mg litre^{-1}, and any increase tends only to increase background fluorescence.

(2) Of the commercially available black membranes the 0·45 μm Sartorius product seems to give the highest counts with the lowest background fluorescence, although their uneven surface can produce focusing problems.

(3) The HBO-200 W mercury pressure illuminator used with BG12 exciter and K480 interference barrier filters, seems to be the best combination for acridine-based fluorochromes.

We have not had the opportunity to try different illuminators, but the use of the K480 interference filter with its steep cut-off instead of the No. 50 barrier filter, would probably decrease background fluorescence.

(4) Yamabe (1973) suggested that when acridine orange combines with double-stranded DNA green fluorescence is produced, and combination with single-strand DNA (denatured) or with RNA produces orange or red fluorescence. If this is so then any acridine-based counting technique could provide the basis for a viable count. Jones (1974) suggests, however, that in practice the differential colour is more a reflection of concentration and contact time than of viability. Jones and Simon (1975) support this by showing that autoclaved bacteria also take up acridine fluorochromes for some considerable time after death, which must cause doubt about its use as a vital stain. Both these two groups of workers and ourselves count only green fluorescent, bacterial-shaped objects with a definite border. This is a subjective assessment, however, and it is probably equally justifiable to count both green and red bacterial-shaped objects.

(5) The count obtained and the distribution of bacteria on the membrane can be affected by the volume filtered.

TABLE 3. The effect of filtration volume on the estimated numbers of epiphytic bacteria obtained by the acridine orange epifluorescence technique

Date	Filtered volume (ml)	Mean bacterial number ± 95% confidence intervals (g^{-1} dry wt. × 10^9)	p
31.10.74	25·0	8·27 ± 1·31	NS
	25·0	7·10 ± 1·22	
	25·0	7·59 ± 1·26	
6.12.74	25·0	5·91 ± 1·13	NS
	2·5	7·60 ± 1·29	
9.1.75	25·0	5·32 ± 1·08	NS
	2·5	7·01 ± 1·24	
7.2.75	25·0	4·01 ± 0·93	0·025
	2·5	7·12 ± 1·25	

Values are means from two membranes ± 95% confidence limits, calculated from untransformed data, assuming a Poisson distribution. The significance of difference between filtration volumes (p) is obtained from a single classification analysis of variance on $\log_{10}$ transformed data.
NS, no significant difference.

Table 3 shows some counts obtained from stomacher-derived epiphyte suspensions, and also shows a comparison between filtering large and small samples. Although in all cases the mean count is higher for a 2·5 ml than for a 25 ml aliquot, in only one of the three comparisons is the difference significant. It also shows that when the sample was counted three times with the same aliquot volume, there was no significant difference between the counts ($p > 0{\cdot}10$). However, filtering smaller volumes markedly reduces background fluorescence and this consequently became part of the standard technique.

The advantages the acridine orange epifluorescence techniques offer are high sensitivity, rapidity, simplicity and low cost per count. They also offer greater specificity than non-fluorochrome methods and the use of membrane filters enables relatively dilute samples to be counted, such as bacteria from individual leaves. The ability of acridine orange to stain debris red or orange is a great advantage with material such as stomacher-derived epiphyte suspensions. In our experience only some small algae show green fluorescence and these are readily distinguished from bacteria by their shape and size. Daley and Hobbie (1975) express the opinion that after appropriate training the technique can be successfully used by unskilled personnel. This is supported by the experience of undergraduate students who successfully used the method after only a short introduction.

Enumeration of Viable Bacteria

The limitations of standard plate counting techniques in the study of microbial ecology are well known to most workers in the field. At present, however, there appear to be no satisfactory substitutes when trying to assess the physiological composition of a natural mixed population. Schmidt (1973) expresses the opinions of a number of workers in stating "the plate count may lead to a complete waste of time and accumulation of useless data." He concludes, however, that plate counts are acceptable when the results are supported by other data, such as direct counts and activity estimates, or when the plating is designed to enumerate specific physiological or other groups of bacteria.

Despite the limitations several groups of workers have estimated viable epiphytic bacterial numbers, and have demonstrated the presence of a substantial population. Potter (1964) found an average viable number of $3{\cdot}84 \times 10^4$ bacteria cm^{-2} on leaves of *Carex* and *Potamogeton*, and Laycock (1974) showed that fronds of *L. longicruris* supported a population of between 8×10^3 and $1{\cdot}5 \times 10^5$ bacteria cm^{-2}. Edwards and Owens (1965) found that on a mixed population of *Elodea*, *Berula* and *Callitriche*, there was an average of 7×10^5 bacteria cm^{-2}, and Ramsay (1974) using *E. canadensis* found an average of 1×10^5 bacteria cm^{-2} using Nutrient Agar (Oxoid).

Media

Jones (1970) and Staples and Fry (1973) have shown that spread-plating bacteria from freshwaters on the Casein-Peptone-Starch (CPS) medium of Collins and Willoughby (1962) produces maximal viable numbers. Strzelczyk and Mielczarek (1971) found the Iron Peptone agar (FePA) of Ferrer *et al.* (1963) to be optimal in isolating bacteria from *E. canadensis*, water and sediment. Work in this laboratory has compared the ability of CPS, FePA and Nutrient Agar (Oxoid) to grow epiphytic bacteria from *E. canadensis*, *Ch. vulgaris* and *P. natans*. In six comparisons highest counts were obtained on CPS, and in most cases these increases were significant at $p = 0{\cdot}01$ (Ramsay and Fry, unpublished).

CPS medium has been used in this laboratory to count viable heterotrophic bacteria from a variety of freshwater macrophytes, and by making additions to the base medium we have enumerated a variety of physiological types. CPS medium contains the following

Soluble Casein (BDH)	0·5 g
Bacto-Peptone (Difco)	0·5 g
Soluble Starch	0·5 g
Glycerol	1·0 ml
$FeCl_3$	4 drops of 0·01% solution
K_2HPO_4	0·2 g
$MgSO_4{\cdot}7H_2O$	0·05 g
Agar	10·0 g
Distilled water	1 litre

Collins *et al.* (1973) describe the preparation procedure, but we have found their technique to be unnecessarily complex. All components except the starch, which needs boiling in a small volume of water, can be dissolved at room temperature with the aid of a magnetic stirrer. The medium is autoclaved at 121° for 15 min and 400 ml used to pour 30 plates in 9 cm petri dishes, and then dried at room temperature for three days before use. The resulting medium should be clear and free from precipitates to facilitate the counting of even the smallest colonies.

Jones (1971) used CPS without its starch and casein to enumerate protease, amylase and lipase exoenzyme producers, by the addition of 0·5% gluten, 0·1% starch and 1% tributyrin respectively. For counting exoenzyme producers on macrophytes we use complete CPS as the base and enumerate protease producers, xylanase producers and paraquat-resistant organisms, by adding 4·5 g litre^{-1} casein, 3 g litre^{-1} xylan (ex larch sawdust, Koch-Light) and 50 mg litre^{-1} Gramoxone S (ICI Plant Protection Ltd). We find no need to add extra starch to count amylase producers. Clear zones develop around xylanase producing colonies, but amylase and protease production is detected by counting clear zones after flooding with 50% Lugol's iodine and 5% trichoracetic acid respectively. Table 4 demonstrates that CPS with added xylan or casein produces heterotroph counts insignificantly different from those on the base CPS medium. This agrees with the findings of Jones (1970) who says that the inorganic components of CPS have a greater effect on numbers than the organic components.

TABLE 4. Mean numbers of viable heterotrophic bacteria epiphytic on seven species of aquatic macrophytes

Number of viable heterotrophs g^{-1} dry weight $\times 10^8 \pm 95\%$ confidence intervals enumerated on		
CPS + 0·5% casein	CPS + 0·3% xylan	CPS
3·41 ± 1·02	2·50 ± 1·01	2·52 ± 0·54

Numbers were means of 26 values obtained in January and February 1975; 95% confidence intervals were calculated on untransformed data.

Counting procedure

Dickinson *et al.* (1975) in an investigation of terrestrial epiphytic bacteria showed that quarter-strength Ringer's buffer was the optimal diluent, and that distilled or tap water were comparatively poor. When enumerating freshwater organisms we use sterile liquid CPS as diluent in preference to sterile fresh water, since the chemical composition of the latter is known to change markedly and sterilizing is likely to have a variable and unpredictable effect on its composition. Liquid CPS also has the properties which have been suggested for a good diluent (Meynell and Meynell, 1965) in containing a buffer, inorganic salts and protein.

Thus the epiphyte suspension is serially diluted in liquid CPS and dilutions spread plated in quadruplicate on each medium. All plates are incubated at 18° for eight days, although work in this laboratory has shown that maximum CPS heterotroph numbers develop after 22 days at 18°. However, incubation for more than eight days allows the exoenzyme zones to merge and become uncountable. Eight-day incubation is therefore used for all media accepting a lower overall count, but enabling the proportion of exoenzyme producers to be estimated. Figure 1 illustrates seasonal changes in bacterial viable numbers for a mixed population of freshwater macrophytes. The total viable heterotrophic number does not seem to vary in any periodic manner, but the exoenzyme producing parts of the population do seem to undergo some regular changes. However, the data are not sufficiently large or regular to enable a rigorous time-series analysis to be applied.

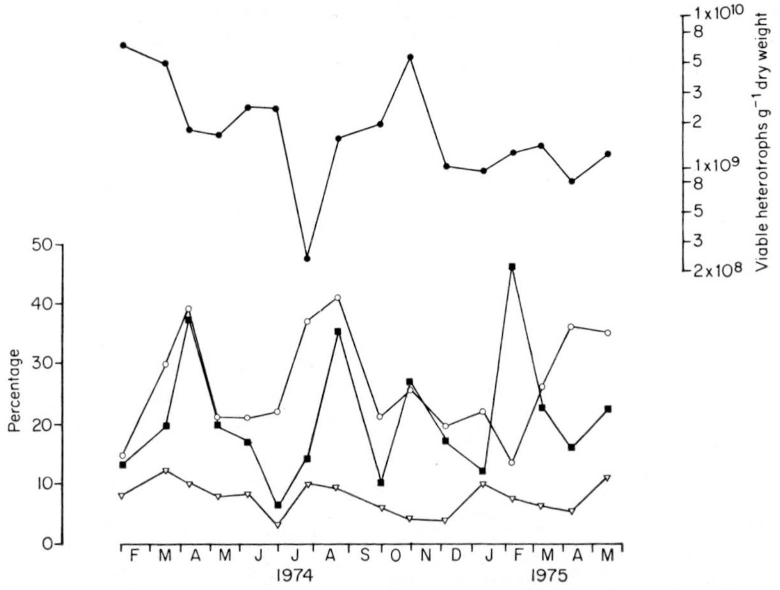

FIG. 1. Changes in numbers and proportions of viable epiphytic bacteria from Coedarhydyglyn pond, between February 1974 and March 1975. Proportions of exoenzyme producers are expressed as a percentage of the viable heterotrophs. ●, heterotrophs; O, protease producers; ■, amylase producers; ▽, xylanase producers.

Assessment of the Activity of Epiphytic Bacteria

Strzelczyk and Mielczarek (1971) compared the activity of bacteria isolated from water, sediment and from *E. canadensis*. Rates of oxygen uptake were measured in a Warburg respirometer using seven substrates: glucose, fructose, sodium acetate, pyruvate, fumarate, gluconate and casamino acids. In all cases epiphytes were more active than planktonic bacteria, which in turn were more active than benthic bacteria.

It was shown by Parsons and Strickland (1962) that the uptake of glucose and acetate by marine bacteria can be described by Michaelis-Menten type kinetics. This has since been shown to be true for organisms in freshwater (Wright and Hobbie, 1966) and their method can be used to estimate and compare the activity of natural populations. A good discussion of the possible limitations of this technique is provided by Wright (1973).

Allen (1971) has used this method to study the heterotrophic uptake of glucose and acetate by bacterial epiphytes from Lawrence Lake, Michigan. He used bacteria isolated from *Chara*, *Scirpus acutus*, *Nuphar* and *Najas flexilis*, and found the average glucose maximum uptake rate to be 10·8 μg glucose $litre^{-1}$ h^{-1} dm^{-2} of leaf surface, and that of acetate to be 21·8 μg acetate $litre^{-1}$ h^{-1} dm^2. He also used plexiglas slides immersed in stands of *Chara* and *Scirpus*, and found that again uptake of acetate was greater than that of glucose.

Activity can similarly be measured by the rate at which ^{14}C-labelled substrate are mineralized to $^{14}CO_2$ (Hobbie and Crawford, 1969). This method has been used to assess bacterial activity in sediments (Harrison *et al.*, 1971) and this is the method we have used to measure epiphytic bacterial heterotrophic activity. It has the advantage over uptake measurement of being extremely simple, rapid and cheap to perform on a large scale, and is therefore ideally suited for use as a routine procedure. Indeed with solid or suspended samples the practical difficulties of separating ^{14}C-labelled bacteria from added ^{14}C-substrate are large, and may never be completely effected since some of the label is probably adsorbed to detritus particles.

General procedure

^{14}C-labelled isotopes D-[U–^{14}C] glucose, [1–^{14}C] glycollic acid and [U–^{14}C] acetic acid (Radiochemical Centre, Amersham) are freeze dried in 4 μCi amounts, and stored in a dessicator at 4°. No observable loss of label occurs within the turnover time of each batch of isotope, bacterial contamination is avoided and reconstitution is rapid. For use, a vial is reconstituted to 1 μCi ml^{-1} with distilled water, and 0·1 ml of this stock added to each test bottle. For routine work four concentrations of substrates are used, each in quadruplicate. One replicate at each substrate concentration is a control, and to this acid is added before addition of the epiphyte suspension. This scheme provides triplicate values at each substrate concentration. Substrate additions are made up of a constant amount of ^{14}C-label (0·1 μCi) and a variable amount of unlabelled substrate, to a final volume of 1 ml. The precise amount of unlabelled substrate added and the range of substrate concentrations possible, depends on the specific activity of the labelled substrate. With high specific activity compounds such as ^{14}C-labelled glucose the minimum concentration is approximately 0·06 μg per bottle, whereas with ^{14}C-labelled glycollate the minimum is more than 1 μg per bottle. Even if a narrow, tenfold range of concentrations is used for the latter type of substrate, the highest concentration is in the range where significant algal heterotrophy can occur (Allen, 1971; Wright and Hobbie, 1966). If this happens the kinetic response deviates from linear at high substrate levels, and approximates to a rectangular hyperbola. However, at concentrations up to 2 mg litre^{-1} of epiphyte suspension, no such deviations have been observed that cannot be explained by normal variability. Thus it seems likely that the heterotrophic activity of epiphyte suspensions is due mainly to bacteria.

Bennett and Hobbie (1972) demonstrated that algae are capable of utilizing, for photosynthesis, $^{14}CO_2$ produced by heterotrophic mineralization of the ^{14}C-substrate. Such a process will lead to an underestimate of bacterial mineralization, but can be inhibited by using dark reaction vessels such as Clinbritic brown glass serum bottles (Malcolm Britton Co. Ltd, 167 Battersea Rise, London SW11). These are closed and made gas tight by rubber serum caps (Malcolm Britton) pierced by pyrex glass filter paper holders (Fig. 2). These are also available commercially (Kontes Glass Co., Vineland, NJ 08360, USA).

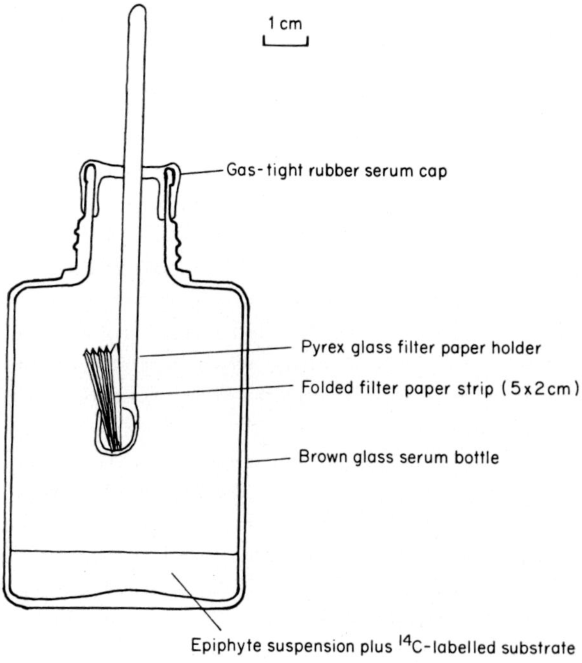

FIG. 2. Apparatus used to collect $^{14}CO_2$ produced during the measurement of heterotrophic activity of epiphytic bacteria.

Five or ten millilitres of epiphyte suspension is added to the substrate in the bottles, and the vessels sealed with the serum cap/filter paper combination. Incubation is at room temperature (20–25°) and the incubation time is chosen so that less than 5% of the total substrate is removed. This ensures that the substrate concentration remains essentially unchanged throughout the experiment. With our samples this is typically between 20 and 60 min.

At the end of the incubation period 0·2 ml of the CO_2 absorbant, 2-phenylethylamine, is added to the filter paper strip through the stopper, and 1 ml of 5 N H_2SO_4 added in the same way to the bottle to stop the reaction and release $^{14}CO_2$. Using 5 N H_2SO_4 is valid only for $^{14}CO_2$ mineralization studies, since Griffiths *et al.* (1974) have shown that acidification releases bound ^{14}C-label when studying the uptake kinetics of glutamic acid.

After an equilibration period of about 30 min, the filter papers are removed into 10 ml of scintillation fluid (6 g 2,5-diphenyloxazole; 0·5 g 1,4-di [2-(5-phenyloxazolyl)] benzene; 100 ml methanol; 900 ml toluene), and the amount of radioactivity counted in a liquid scintillation counter. In the Beckman LS 235 machine counting efficiency was found to be about 90% by an external standard method, and the difference between counts per minute and counts corrected to disintegrations per minute, was insignificant ($p \gg 0{\cdot}10$).

The fraction t/f (incubation time in hours/fraction of added ^{14}C appearing as $^{14}CO_2$) when plotted against A (added substrate concentration) should yield a straight line with positive gradient and *y* intercept if the Michaelis-Menten kinetic model is true. The significance of linear regression and the best line to fit the data is calculated by an analysis of variance procedure (Sokal and Rohlf, 1969). From this line it is possible to calculate three parameters which can provide some measure of heterotrophic activity (for details see Table 5).

(1) V_m, maximum velocity—the maximum possible rate of mineralization of substrate to CO_2: not to be confused with the natural velocity V, which is the actual rate of mineralization occurring in the sample.

(2) T, turnover time—the time in hours for the natural substrate concentration Sn to be mineralized to CO_2 at the natural velocity, V.

(3) K + Sn—a combined term of K a constant, which is probably an affinity or a transport constant (see comment following Crawford *et al.*, 1973), and Sn the natural substrate concentration.

TABLE 5. ^{14}C-substrate mineralization kinetic equations

Basic kinetic equation	$\frac{S}{V} = \frac{K}{V_m} + \frac{S}{V_m}$ . . . . (1)
Found by experiment	$V = \frac{f\,(Sn + A)}{t}$ $\therefore \frac{t}{f} = \frac{(Sn + A)}{V}$ . . . . . . (2)
	Since $S = Sn + A$ . . . (3) then $\frac{t}{f} = \frac{S}{V}$ . . . (4)
By substitution	(3) into (1) $\frac{S}{V} = \frac{K}{V_m} + \frac{(Sn + A)}{V_m}$ . . . . (5)
	(4) into (5) $\frac{t}{f} = \frac{K}{V_m} + \frac{(Sn + A)}{V_m}$
	$\therefore \frac{t}{f} = \frac{K + Sn}{V_m} + \frac{1}{V_m} \, . \, A$
By experiment	find t/f values for different values of A Plot t/f against A, then (i) The reciprocal of the gradient of the line $= V_m$ μg unit^{-1} h^{-1} (ii) The y intercept divided by $V_m = K + Sn$ μg unit^{-1} (iii) When $A = O$, $\frac{t}{f} = \frac{Sn}{V}$ (2) and $\frac{Sn}{V} = \frac{K + Sn}{V_m} = T\,h^{-1}$ thus the y intercept when $A = O$ is the turnover time $T\,h^{-1}$

V, reaction velocity; S, substrate concentration (total); V_m, maximum reaction velocity; K, constant, $= S$ for $V_m/2$; f, fraction of added isotope appearing as $^{14}CO_2$; t, incubation time (h); Sn, natural substrate concentration; A, added substrate concentration.

Preparation of epiphyte suspensions for activity measurement

In most cases, as in natural waters, Sn is much less than A, and t/f is directly proportional to A. If, however, Sn becomes large with respect to A, t/f tends to become independent of A and a plot of t/f against A produces a line with a gradient insignificantly different from zero or even negative. This occurs with stomacher-derived epiphyte suspensions, since much intracellular plant material is released by mechanical damage, and makes the apparent Sn large. There are at least three ways of approaching this problem.

By using artificial substrata

Allen (1971) has used plexiglas slides to simulate the surface of aquatic macrophytes, because he found that epiphyte suspensions made from swabbings of macrophytes were contaminated with plant material. It is doubtful, however, whether an inert material like plexiglas can adequately mimic a dynamic environment such as the surface of a plant, especially when measuring an activity parameter.

By using whole plant parts

In this method plant parts such as whole leaves are put into bottles with distilled water instead of an epiphyte suspension. This is only reliable if bacterial epiphytes alone take up and mineralize the substrate. Using the autoradiographic method of Ramsay (1974), it has been shown in this laboratory (Ramsay and Fry, unpublished) that when epiphytes are exposed to low concentrations (50 μg litre^{-1}) of ^{3}H-glucose, only the bacteria become tritiated and that epiphytic algae and plant material were never labelled. Moreover, if epiphytes were removed from *Elodea* leaves by scraping, at least 67% of the glucose taken up by intact leaves was accounted for by the removed epiphytes. The activity remaining on the leaves was probably due to bacteria not removed by the scraping procedure, since bacteria could still be seen by direct observation of the scraped leaves.

These results indicate that the measurement of glucose uptake and mineralization using whole plant parts and low substrate concentrations reflects epiphytic bacterial activity. The major disadvantage of this method however is the high degree of variability and subsequent low number of significant regression lines obtained. Of 35 experiments performed in 34 there was no significant difference ($p > 0{\cdot}05$) in the t/f values with change in A. Moreover, only 17% of the experiments produced significant linear regression at the $p = 0{\cdot}05$ level, and only 28·6% were significant at $p = 0{\cdot}10$. Also 26% of lines gave negative slopes and therefore negative activity values. However, it is only with methods such as this, which measure epiphytic activity on the plants that true measures of the actual bacterial activity can be made. Only with these methods is the natural substrate concentration retained, and since both measures of turnover time and K are directly related to Sn only with such methods can ecologically meaningful estimates of T or K + Sn be made.

By washing epiphytes free of substrates

Because of the variability problem associated with the previous method we have devised a "washing" method to assess the activity of epiphytes

TABLE 6. Distribution of carbohydrate and protein in centrifuged epiphyte fractions

Fraction	Total carbohydrate*			Protein†		
	$\mu g\ g^{-1}$	± 95% confidence intervals	Concentration as percentage of initial suspension	$\mu g\ g^{-1}$	± 95% confidence intervals	Concentration as percentage of initial suspension
Initial epiphyte suspension	34·20	5·84	100	35·70	9·11	100
Final epiphyte suspension	13·84	1·74	40·5	7·68	2·75	21·5
First supernatant	17·30	4·10	50·6	26·58	6·62	74·5
Second supernatant	0·75	0·09	2·20	1·32	0·43	3·70
Third supernatant	0·33	0·14	0·97	0·49	0·51	1·40

* Total carbohydrate estimated by the method of Gerchakov and Hatcher (1972).
† Protein estimated by the method of Lowry *et al.* (1951).
Values are means of 6 (carbohydrate) or 4 (protein).

removed by stomaching. The procedure is to centrifuge 400 ml of stomacher–derived epiphyte suspension at 10 000 *g* for 15 min at 20°, and resuspend the pellet in 400 ml of sterile distilled water. This is done a total of three times and the final pellet resuspension used in the activity estimation. Table 6 shows that the soluble protein and total carbohydrate concentrations in the final supernatant are reduced by at least 55 times to about 1% of the original total value. The amount of these materials in the final resuspension which are soluble, and hence freely available, is therefore probably less than 1%. Table 7 shows that the centrifugation procedure has no significant effect on bacterial viability

TABLE 7. Effect of three centrifugations at 10 000 *g* on the viability of epiphytic bacteria

		Viable number of bacteria g^{-1} dry weight ± 95% confidence intervals in	
Date	Organisms	Initial suspension	Final suspension
14.3.74	Heterotrophs	$4{\cdot}84 \pm 1{\cdot}56 \times 10^9$	$4{\cdot}40 \pm 1{\cdot}09 \times 10^9$
	Xylanase producers	$5{\cdot}90 \pm 1{\cdot}34 \times 10^8$	$4{\cdot}72 \pm 1{\cdot}36 \times 10^8$
	Amylase producers	$9{\cdot}15 \pm 2{\cdot}06 \times 10^8$	$7{\cdot}68 \pm 3{\cdot}14 \times 10^8$
4.4.74	Heterotrophs	$1{\cdot}82 \pm 0{\cdot}59 \times 10^9$	$1{\cdot}82 \pm 0{\cdot}64 \times 10^9$
	Xylanase producers	$1{\cdot}82 \pm 1{\cdot}01 \times 10^8$	$1{\cdot}35 \pm 0{\cdot}85 \times 10^8$
	Amylase producers	$6{\cdot}79 \pm 2{\cdot}46 \times 10^8$	$5{\cdot}77 \pm 4{\cdot}89 \times 10^8$

Values are means of 4 counts, 95% confidence intervals calculated on untransformed data.
The final suspension is the pellet from the initial suspension, washed by three centrifugations at 10 000 *g*, resuspended in distilled water.

($p > 0{\cdot}10$), and observation of the final pellet resuspension shows many bacteria to be motile. It is therefore presumed that the centrifugation procedure has little effect on bacterial activity.

Typical kinetic plots obtained before and after centrifugation are shown in Figs 3 and 4, which also show the inhibitory effect of adding the first supernatant which contains most of the soluble material, to the final pellet resuspension. The value of glucose V_m for the initial suspension could not be calculated due to its negative slope, but the final pellet resuspension had a V_m of 0·490 μg glucose g^{-1} dry weight of plant h^{-1}. Moreover, the regression line produced by this pellet resuspension was very significant ($p < 0{\cdot}025$). If the removed substrate represented by the first supernatant is added to the final resuspension, it should inhibit the activity shown by the final resuspension alone. Indeed the V_m decreases to 0·203 μg glucose g^{-1} h^{-1}, and the regression is not significant ($p > 0{\cdot}10$). These findings support the predictions of the inhibitory effect of a large apparent Sn, and show that measures of activity in the final pellet resuspension give reliable kinetics. Using this method, in 50 experiments performed with ^{14}C-labelled glucose over a 59 week period, only 3 (6%) showed no significant linear regression ($p > 0{\cdot}10$). Thirty-eight (76%) were significant ($p < 0{\cdot}05$) and there were no negative slopes. The disadvantage of this type of method is that the activity parameters, especially

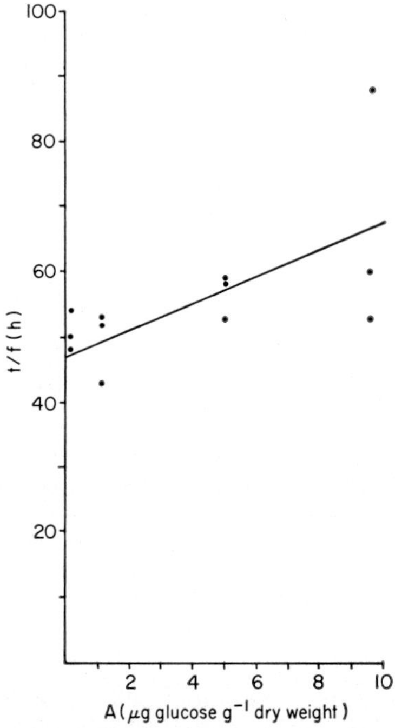

FIG. 3. Kinetic plot of mineralization of ^{14}C-labelled glucose to $^{14}CO_2$ by epiphytic bacteria in the final resuspension, obtained after three centrifugations, of stomacher derived epiphyte suspension. The plot is of t/f against A, and the line is the least squares best fit obtained by analysis of variance. The regression equation for this line is $y = 2{\cdot}04x + 47{\cdot}1$, and linear regression is significant ($p < 0{\cdot}025$). The kinetic constants are: $V_m = 0{\cdot}47\ \mu g\ g^{-1}\ h^{-1}$, $T = 47{\cdot}1$ h, $K + Sn = 23{\cdot}1\ \mu g\ g^{-1}$.

T and K + Sn, bear no relation to their values *in situ*. Assuming that the metabolic state of the organisms is not widely affected, V_m measured in this way is probably representative of V_m *in situ* since, unlike K + Sn or T it is a property of the epiphytes only, and not dependent on V or Sn. However, although the method does not provide activity estimates of bacteria on the plant, the more reliable kinetics make it a useful comparative method when, for example, studying changes. Table 8 shows a

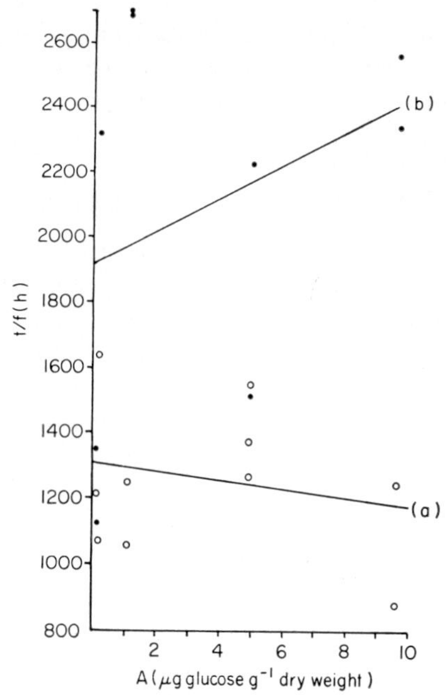

FIG. 4. Kinetic plot of mineralization of ^{14}C-labelled glucose to $^{14}CO_2$ by epiphytic bacteria from (a) the original stomacher derived epiphyte suspension (○), and (b) a reconstitution of the pellet obtained after three centrifugations and the first supernatant (●). The plots are of t/f against A, and the lines are the least squares best fit obtained by analysis of variance. The regression equation is (a) $y = 14{\cdot}4x + 1309$ and linear regression is not significant ($p > 0{\cdot}10$), (b) $y = 49{\cdot}2x + 1923$ and linear regression is not significant ($p > 0{\cdot}10$). For line (a) no meaningful kinetic constants can be calculated due to its negative slope. The kinetic constants for line (b) are: $V_m = 0{\cdot}020\ \mu g\ g^{-1}\ h^{-1}$, $T = 1923$ h, $K + Sn = 39{\cdot}1\ \mu g\ g^{-1}$.

TABLE 8. Comparison of the heterotrophic activity of planktonic, benthic and epiphytic bacteria from Coedarhydyglyn pond

Sample	V_m for mineralization of substrate to CO_2, $\mu g \times 10^9$ bacteria* h^{-1} Glucose	Glycollate	Acetate
Water	0·3662	0·6029	0·2577
Macrophyte	0·4846	0·4507	0·2647
Sediment	0·0172	0·0132	0·0122

* Numbers of bacteria obtained in June 1976 were measured by an acridine orange epifluorescence technique.

comparison of heterotrophic activities measured by this technique of planktonic, benthic and epiphytic bacteria from a freshwater pond. It is noteworthy that epiphytic bacteria are at least as active as planktonic bacteria, and apparently more active towards glucose and acetate, at least in early summer.

Assessment of Proportions of Active Epiphytic Bacteria

An autoradiographic method has been described (Ramsay, 1974) for assessing the proportion of active epiphytic bacteria, that is those able to take up D-[6–^{3}H]-glucose. Although the details of the technique have been fully described by Ramsay an outline of the method will be given here. Small pieces of plant material are incubated with 20 μCi ml^{-1} of ^{3}H-labelled glucose for 2 h before fixing with formalin. Epiphytes are removed from the plant surface by scraping; a film of them is made on a microscope slide which is dipped in photographic emulsion. The emulsion is exposed to the radiation in the epiphyte film for seven days and the slides processed by the technique of Caro and van Tubergen (1962). The developed film is then stained, a permanent mount prepared and at least 400 bacteria scored as labelled or unlabelled by bright field microscopy. The proportion of active bacteria can then be calculated.

Few studies have as yet used this technique. However, in one study (Ramsay and Fry, 1976) a large proportion of epiphytic bacteria appeared to take up glucose. In this case although the proportions obtained were very variable about 60% of the bacterial epiphytes on *E. canadensis* and about 80% of those on *C. vulgaris* were able to take up glucose.

Acknowledgements

Most of this work was carried out during the tenure of a Natural Environment Research Council research grant for which we are grateful. Some of the results were obtained by Mr. A. Johnson and most of the experiments were carried out with macrophytes from a small pond at Coedarhydyglyn, Cardiff; permission to sample this was kindly given by Sir Cenydd Traherne.

References

ADAMS, M. S., TITUS, J. & MCCRACKEN, M. (1974). Depth distribution of photosynthetic activity in a *Myriophyllum spicatum* community in Lake Wingra. *Limnol. Oceanogr.*, **19**, 377.

ALLANSON, B. R. (1973). The fine structure of the periphyton of *Chara* sp. and *Potamogeton natans* from Wytham Pond, Oxford, and its significance to the macrophyte-periphyton metabolic model of R. G. Wetzel and H. L. Allen. *Freshwat. Biol.*, **3**, 535.

ALLEN, H. L. (1971). Primary productivity, chemo-organotrophy and nutritional interactions of epiphytic algae and bacteria in macrophytes in the littoral of a lake. *Ecol. Monogr.*, **41**, 97.

BENNETT, M. E. & HOBBIE, J. E. (1972). The uptake of glucose by *Chlamydomonas* sp., *J. Phycol.*, **8**, 392.

BERMAN, T. & HOLM-HANSEN, O. (1974). Release of photoassimilated carbon as dissolved organic matter by marine phytoplankton. *Mar. Biol.*, **28**, 305.

CARO, L. G. & VAN TUBERGEN, R. P. (1962). High resolution autoradiography 1. Methods. *J. Cell. Biol.*, **15**, 173.

CHAN, E. C. S. & MCMANUS, E. A. (1967). Development of a method for the total count of marine bacteria on algae. *Can. J. Microbiol.*, **13**, 295.

COLER, E. A. & GUNNER, H. B. (1969). The rhizosphere of an aquatic plant *Lemna minor*. *Can. J. Microbiol.*, **15**, 964.

COLLINS, V. G. & WILLOUGHBY, L. G. (1962). The distribution of bacteria and fungal spores in Blelham Tarn with particular reference to an experimental overturn. *Arch. Mikrobiol.*, **43**, 294.

COLLINS, V. G., JONES, J. G., HENDRIE, M. S., SHEWAN, J. M., WYNN-WILLIAMS, D. D. & RHODES, M. E. (1973). Sampling and estimation of bacterial populations in the aquatic environment. In *Sampling—microbiological monitoring of environments* (Board, R. G. & Lovelock, D. W., eds). Soc. appl. Bact. Tech. Ser. No. 7. London and New York: Academic Press, p. 77.

CRAWFORD, C. C., HOBBIE, J. E. & WEBB, K. L. (1973). Utilization of dissolved organic compounds in an estuary. In *Estuarine microbial ecology*, No. 1 (Stevenson, L. H. & Colwell, R. R., eds). The Belle W. Barusch Library in Marine Science, Univ. South Carolina Press, p. 169.

DALEY, R. J. & HOBBIE, J. E. (1975). Direct counts of aquatic bacteria by a modified epifluorescence technique. *Limnol. Oceanogr.*, **20**, 875.

DICKINSON, C. H., AUSTIN, B. & GOODFELLOW, M. (1975). Quantitative and qualitative studies of phylloplane bacteria from *Lolium perenne*. *J. gen. Microbiol.*, **91**, 157.

EDWARDS, R. W. & OWENS, M. (1960). The effects of plants on river conditions. I. Summer crops and estimates of net productivity of macrophytes in a chalk stream. *J. Ecol.*, **48**, 151.

EDWARDS, R. W. & OWENS, M. (1965). The oxygen balance in streams. In *Ecology and the industrial society* (Goodman, G. T., Edwards, R. W. & Lambert, J. M., eds). Oxford: Blackwell, p. 149.

ELLIOT, J. M. (1971). Some methods for the statistical analysis of samples of benthic invertebrates. *FBA Scientific Publication No.* **25**.

FENCHEL, T. (1970). Studies on the decomposition of organic detritus from the turtle grass *Thalassia testuclinum*. *Limnol. Oceanogr.*, **15**, 14.

FERRER, E. B., STAPERT, E. M. & SOKOLSKI, W. T. (1963). A medium for improved recovery of bacteria from water. *Can. J. Microbiol.*, **9**, 420.

FRY, J. C. & RAMSAY, A. J. (1977). Changes in the activity of epiphytic bacteria of *Elodea canadensis* and *Chara vulgaris* following treatment with the herbicide, paraquat. *Limnol. Oceanogr.*, **22**, 556.

GERCHAKOV, S. M. & HATCHER, P. G. (1972). Improved technique for the analysis of carbohydrates in sediments. *Limnol. Oceanogr.*, **17**, 938.

GILLESPIE, D. M. & BROWN, C. J. D. (1966). A quantitative sampler for macroinvertebrates associated with aquatic macrophytes. *Limnol. Oceanogr.*, **11**, 404.

GRIFFITHS, R. P., HANUS, F. J. & MORITA, R. Y. (1974). The effects of water-sample treatment on the apparent uptake of glutamic acid by natural marine microbial populations. *Can. J. Microbiol.*, **20**, 1261.

Harrison, M. J., Wright, R. T. & Morita, R. Y. (1971). Method for measuring mineralisation in lake sediments. *Appl. Microbiol.*, **21**, 698.

Harrod, J. J. & Hall, R. E. (1962). A method for determining the surface areas of various aquatic plants. *Hydrobiologia*, **20**, 173.

Hobbie, J. E. & Crawford, C. C. (1969). Respiration corrections for bacterial uptake of dissolved organic compounds in natural waters. *Limnol. Oceanogr.*, **14**, 528.

Hossell, J. C. & Baker, J. H. (1976). The distribution and characterisation of bacteria on the surfaces of some river macrophytes. *J. appl. Bact.*, **41**, 14.

Hossell, J. C. & Baker, J. H. (in press). A note on the enumeration of epiphytic bacteria by microscopic methods with particular reference to two freshwater plants. *J. appl. Bact.*

Hough, R. A. & Wetzel, R. G. (1975). The release of dissolved organic carbon from submersed macrophytes. Diel, seasonal and community relationships. *Verh. int. Verein. theor. angew Limnol.*, **19**, 939.

Jones, J. G. (1970). Studies on freshwater bacteria. Effect of medium composition and method on estimates of bacterial populations. *J. appl. Bact.*, **33**, 679.

Jones, J. G. (1971). Studies on Freshwater Bacteria: Factors which influence the population and its activity. *J. Ecol.*, **59**, 593.

Jones, J. G. (1974). Some observations on direct counts of freshwater bacteria obtained with a fluorescence microscope. *Limnol. Oceanogr.*, **19**, 540.

Jones, P. C. T. & Mollison, J. E. (1948). A technique for the quantitative estimation of soil microorganisms. *J. gen. Microbiol.*, **2**, 54.

Jones, J. G. & Simon, B. M. (1975). An investigation of errors in direct counts of aquatic bacteria by epifluorescence microscopy, with reference to a new method for dyeing membrane filters. *J. appl. Bact.*, **39**, 317.

Koegel, R. G., Bruhn, H. D. & Livermore, D. F. (1972). Improving surface water conditions through control and disposal of aquatic vegetation. *Univ. Wisconsin Water Resour. cent. Rep. OWRR–B–018–WIS.* Madison, p. 46.

Laycock, R. A. (1974). Bacteria associated with the surface of *Laminaria* fronds. *Mar. Biol.*, **25**, 223.

Lowry, A. H., Rosebrough, N. J., Farr, A. L. & Randall, R. J. (1951). Protein measurement with the Folin phenol reagent. *J. biol. Chem.*, **193**, 265.

Meynell, G. G. & Meynell, E. (1965). *Theory and practice in experimental bacteriology*. Cambridge: Cambridge Univ. Press.

Parsons, T. R. & Strickland, J. D. H. (1962). On the production of particulate organic carbon by heterotrophic processes in sea water. *Deep-Sea Res.*, **8**, 211.

Potter, L. F. (1964). Planktonic and benthic bacteria of lakes and ponds. In *Principles and applications in aquatic microbiology* (Heukelekian, H. & Dondero, N. C., eds). New York: Wiley, p. 148.

Ramsay, A. J. (1974). The use of autoradiography to determine the proportion of bacteria metabolising in an aquatic habitat. *J. gen. Microbiol.*, **80**, 363.

Ramsay, A. J. & Fry, J. C. (1976). Response of epiphytic bacteria to the treatment of two aquatic macrophytes with the herbicide paraquat. *Water Res.*, **10**, 453.

Rich, P. H., Wetzel, R. G. & Thuy, N. V. (1971). Distribution, production and role of aquatic macrophytes in a southern Michigan marl lake. *Freshwat. Biol.*, **1**, 3.

Rosine, W. N. (1955). The distribution of invertebrates on submerged aquatic plant surfaces in Muskee Lake, Colorado. *Ecology*, **36**, 308.

Schmidt, E. L. (1973). Panel Discussion 1. The traditional plate count technique among modern methods. In *Modern methods in the study of microbial ecology* (Roswall, T., ed.). Bulletins from the ecological research committee 17. Swedish Natural Science Research Council, p. 453.

SCULTHORPE, C. D. (1967). *The biology of aquatic vascular plants.* London: Edward Arnold, p. 610.

SHARPE, A. N. (1973). Automation and instrumentation developments for the bacteriology laboratory. In *Sampling—microbiological monitoring of environments* (Board, R. G. & Lovelock, D. W., eds). Soc. appl. Bact. Tech. Ser. No. 7. London and New York: Academic Press, p. 197.

SHARPE, A. N. & JACKSON, A. K. (1972). Stomaching: a new concept in bacteriological sample preparation. *Appl. Microbiol.*, **24**, 175.

SOKAL, R. R. & ROHLF, F. J. (1969). *Biometry*. San Francisco: W. H. Freeman.

SOROKIN, Y. I. & OVERBECK, J. (1972). The determination of microbial numbers and biomass. In *IBP handbook 23, microbial production and decomposition in freshwater* (Sorokin, Y. I. & Kadota, H., eds). London: Blackwell, p. 40.

STAPLES, D. G. & FRY, J. C. (1973). A medium for counting aquatic heterotrophic bacteria in polluted and unpolluted waters. *J. appl. Bact.*, **36**, 179.

STRZELCZYK, K. E. & MIELCZAREK, A. (1971). Comparative studies of planktonic, benthic and epiphytic bacteria. *Hydrobiologia*, **38**, 67.

WESTLAKE, D. F. (1968). The weight of water weed in the River Frome. *River Bds Ass. Yb.*, 1968, 59.

WETZEL, R. G. (1969). Excretion of dissolved organic compounds by aquatic macrophytes. *Bioscience*, **19**, 539.

WETZEL, R. G. (1975). *Limnology*. Philadelphia: W. B. Saunders, p. 743.

WETZEL, R. G. & ALLEN, H. L. (1972). Functions and interactions of dissolved organic matter and the littoral zone in lake metabolism and eutrophication. In *Productivity problems of freshwaters* (Kajek, Z. & Hillbricht-Ilkowska, A., eds). Warszawa-Krakow: PWN Polish Scientific Publishers, p. 333.

WRIGHT, R. T. (1973). Some difficulties in using ^{14}C organic solutes to measure heterotrophic bacterial activity. In *Estuarine microbial ecology*, No. 1 (Stevenson, L. H. & Colwell, R. R., eds). The Belle W. Barusch Library in Marine Science, Univ. South Carolina Press, p. 199.

WRIGHT, R. T. & HOBBIE, J. E. (1966). Use of glucose and acetate by bacteria and algae in aquatic ecosystems. *Ecology*, **47**, 447.

YAMABE, S. (1973). Further spectophotometric studies on the binding of acridine orange with DNA. *Arch. Biochem. Biophys*, **154**, 19.

Methods for Studying Micro-organisms in Decaying Leaves and Wood in Freshwater

L. G. WILLOUGHBY

Freshwater Biological Association, Ambleside, Cumbria, England

Introduction

With the increasing pollution of the marine environment and the consequent re-examination of the potentiality of freshwater as a human food resource there is accelerating interest in the food chains which culminate in fish production. Imported or allochthonous materials derived from terrestrial vegetation, such as the leaves and wood of trees, find their way into water with surprising frequency and in some bodies of freshwater contribute a major part of the total energy budget. In extreme upland situations tree material is often less conspicuous, or even absent, but here blown-in grass from the surrounding moors may be present in quantity (see Willoughby and Minshall, 1975). All these allochthonous materials, especially shed leaves, are consumed by invertebrate animals which in their turn are preyed on by the resident fish. Microbial decay proceeds simultaneously and it has come to be realized that such "conditioning" often renders the vegetable food more palatable and nutritious. In this chapter decaying aquatic macrophytes which clearly also contribute to the food chains are not considered.

In consuming decaying leaves the invertebrates also consume the resident microflora and it has been suggested that the latter makes a direct contribution to the diet as a highly nutritious additive (Bärlocher and Kendrick, 1975). If this is so the total amount of fungus, actinomycete or bacterial biomass which is present is the important thing to study and qualitative distinctions will be necessary if different taxa have differing food values. On the other hand, there is also the possibility that the beneficial effect of microbes is exerted indirectly, by partial or complete hydrolysis of high polymers in the leaf material which are thus rendered more digestible and nutritious. There seems to be little or no information on this aspect. However, this particular contribution is less concerned with the wider implications than in the actual microbial colonizations and the methods used to examine them.

Fungi

The fungal colonization of leaves in streams can be studied by exposing them *in situ* in nylon net bags or in open-ended tubes. From many points of view it seems desirable to use the leaf material in the natural shed condition (unsterilized) and accordingly dry leaf collections must be made in the autumn and stored for subsequent use throughout the year if the research programme demands this. Experience has shown that leaves strung by the petiole in protective tubes (Fig. 1) allow a more

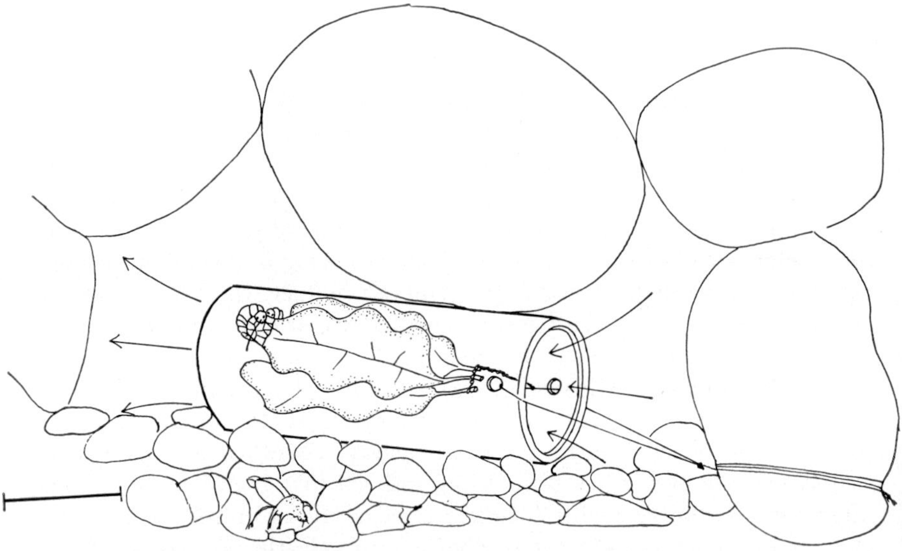

FIG. 1. Exposure of oak leaves to stream colonization in a protective tube. The scale line represents 5 cm.

luxuriant fungal colonization than leaves left compacted in bags, but it could be argued that both of these methods of exposure are "unnatural". Leaves can also be exposed on mesh trays enclosed in perforated boxes but a reduced water flow across the material might be a disadvantage here. The exposure of wood to colonization has been made by enclosing samples in Tygon Screen cloth bags of 2 mm mesh, sewn up with nylon thread and stapled (Willoughby and Archer, 1973).

Recovered leaves, either from experimental exposures or from natural stream collections, can be examined directly for fungi or induced to produce more spores for identification in laboratory incubation (see below). Unorthodox manipulation of leaf material in the laboratory can be productive of hitherto unsuspected forms. For example when recovered alder and oak leaves were cut into squares and incubated in water at 25° the very rare chytrid *Saccopodium*, unrecorded since 1877, grew out from the cut edges (Willoughby, 1966). This evocation may be attributed to severing and stimulating mycelium buried in the substratum. Blending of freshly recovered leaves in sterile water, taking care to avoid excess heat and plating on to 1% malt-penicillin-streptomycin agar is productive of growing fungi but there are limitations in this method for comparative enumeration. Malt agar is highly acid (pH 4·9) but experience shows that it is suitable for the majority of leaf fungi and is not unduly selective. A more serious disadvantage is that the aquatic hyphomycetes, the predominant leaf colonists, do not sporulate directly on the growth medium and laborious cutting out of colonies into water is necessary to induce spores and allow identification. Again, rapidly growing colonies of *Aureobasidium pullulans* or of *Phoma* spp. are often present and these can vitiate realistic counts (Fig. 2a). On the credit side however the blending and plating method does draw attention to the unicellular yeasts which are always represented. A typical 0·25 ml aliquot of a 70 ml blend derived from an oak leaf exposed for 72 days in a stream yielded 1 *Penicillium*, 2 *Aureobasidium pullulans*, 2 unicellular yeasts, 34 aquatic hyphomycetes and 34 unidentified colonies. The aquatic hyphomycetes comprised *Alatospora acuminata*, *Anguillospora longissima*, *Clavariopsis aquatica*, *Lemonniera aquatica* and *Lunulospora curvula*.

Methods of vigorous blending aimed at giving a spatial separation of the leaf substratum and its micro-organisms have the disadvantage that they may eventually destroy the mycelial component. For example Johnston (1972) showed that when spores and mycelium of the actinomycete *Micromonospora* were blended the viable count from plating at first increased, and then decreased to a constant level. This was attributed to an initial spatial separation of the mycelial component followed by its eventual destruction, leaving only residual viable spores. A method for the quantitative assessment of aquatic hyphomycete fungi on leaves which avoids potentially destructive procedures and aims at keeping the substratum and its associated microflora intact has been investigated. The method is based on stimulation of the substrate mycelium to realize its maximum spore potential. The underlying observations are that although freshly collected leaf material may show prolific spore production (Fig. 3b), this is a fairly rare condition. A sparse spore production, or even a complete absence of spores may be encountered, even though mycelium is present (Fig. 3a). It is known that if this material is kept in

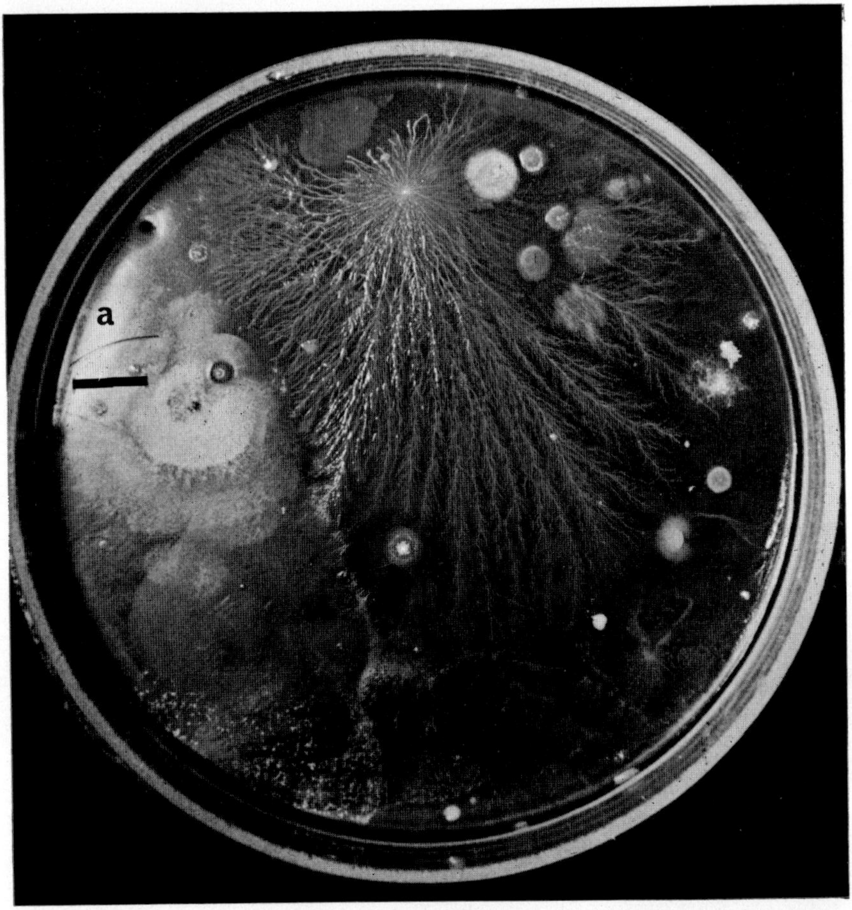

FIG. 2. (a) Fungi. Leaf blend plated on malt-penicillin-streptomycin agar for fungi showing a large colony of *Aureobasidium pullulans* spreading from the top and smaller colonies of aquatic hyphomycetes. The scale line represents 1 cm.

shallow water in the laboratory overnight or longer, spore production and release is tremendously accelerated. The reason for this is not clear and although it may be regarded as an "enrichment" phenomenon there is also the possibility that the breaking of some kind of natural stasis is involved. The site is visited with sterile plastic bags (17·5 × 30 cm) for subsequent use in a Stomacher 400 (A. J. Seward Company), each containing 80 ml of sterile distilled water. Individual leaves are removed to separate bags which, on return to the laboratory, are opened and placed in a shallow tray incubated at 17°. Under these conditions 40 h appears

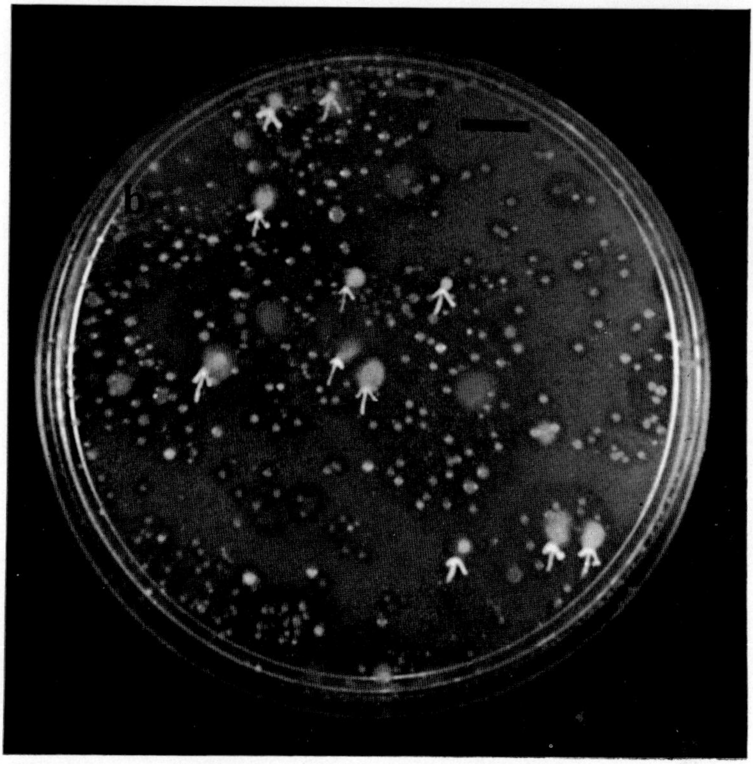

FIG. 2. (b) Actinomycetes. Leaf wash plated on chitin-actidione agar for actinomycetes showing clearing of the medium by actinomycetes and bacteria. *Actinoplanes* colonies marked with arrows. The scale line represents 1 cm.

to be sufficient time for maximum spore production to be realized. Each bag is given 30 s gentle paddle agitation with a Stomacher 400 and 1 ml aliquots of the water are transferred to a Sedgwick-Rafter counting chamber, where the spores are identified (Ingold, 1975) and counted. An alternative procedure here might be to membrane-filter the water and count the spores on the membrane (see Iqbal and Webster, 1973). Experience has shown that the number of spores recovered is not increased by longer periods of paddle agitation up to five minutes. Where spores are present in quantity there is good random distribution of these over the one microlitre grid units of the Sedgwick-Rafter counting chamber and there is very good replication between counts made on successive 1 ml samples (Fig. 4a). Between successive samples it is advisable to clean the counting chamber with detergent, taking care not to scratch the plastic grid area, to remove any adherent spores. Where spores are very

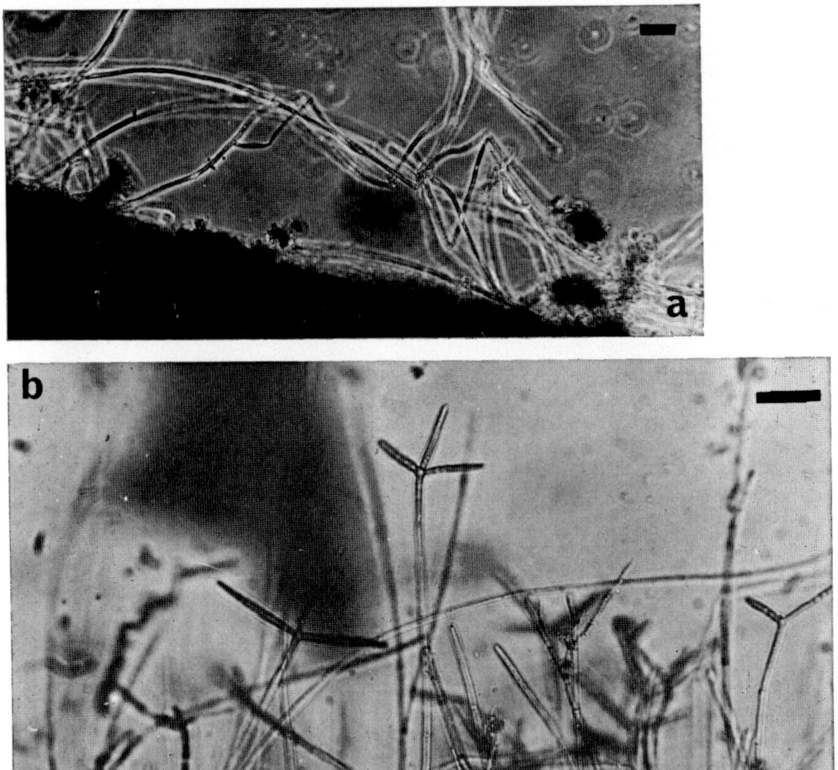

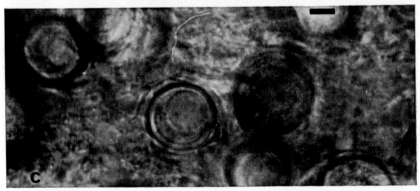

FIG. 3. Fungi. (a) Degraded oak leaf from a stream collection showing sterile mycelium; (b) allochthonous grass collected from Hardknott Gill, River Duddon, showing mycelium and spores of *Tricladium giganteum*; (c) *Pythium* mycelium and oospores growing on cellophane deposited in a nutrient-rich stream. The scale line represent 50 μm for (a) and (b); 5 μm for (c).

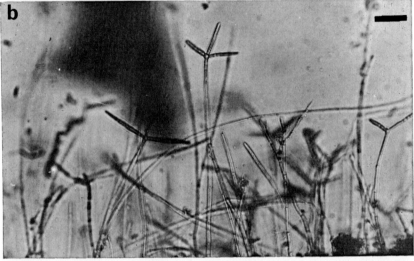

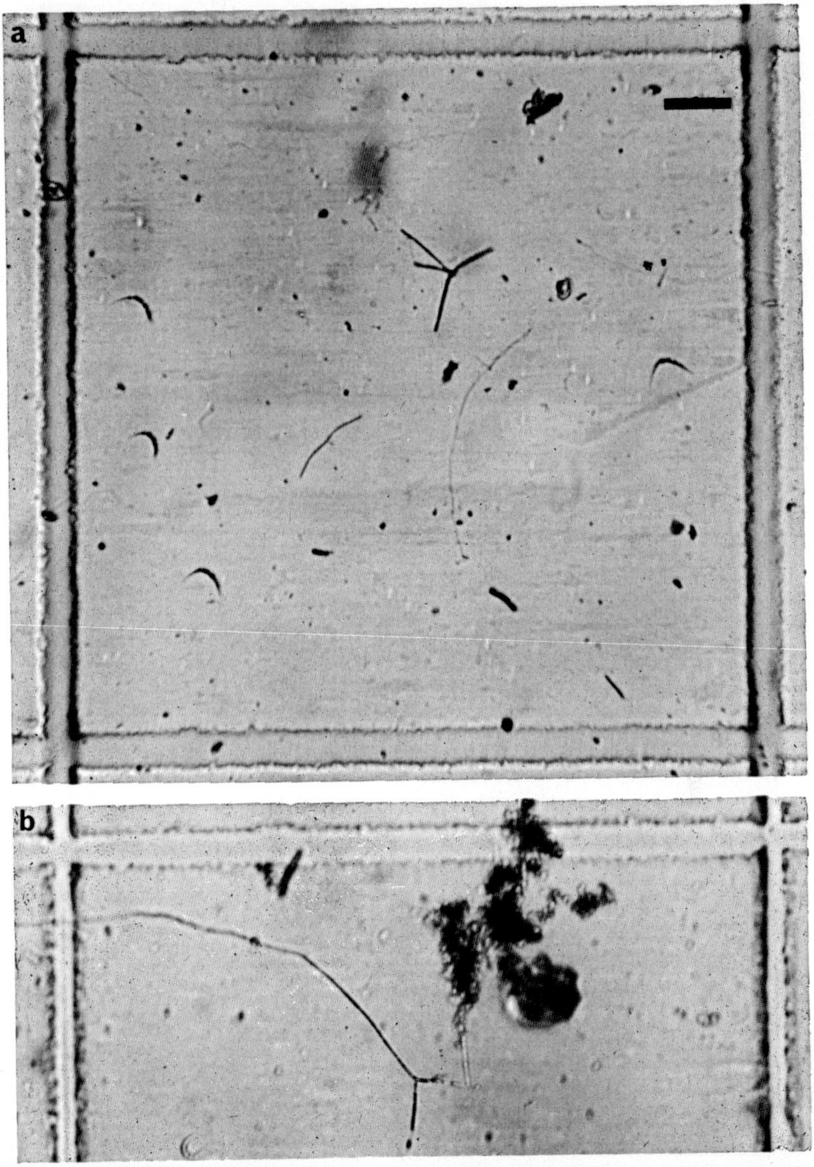

FIG. 4. Fungi. (a) Microlitre grid unit in a Sedgwick-Rafter counting chamber showing four spores of *Anguillospora crassa* and one of *Tetrachaetum elegans*; (b) mycelium making a grid-line interception. The scale line represents 100 μm.

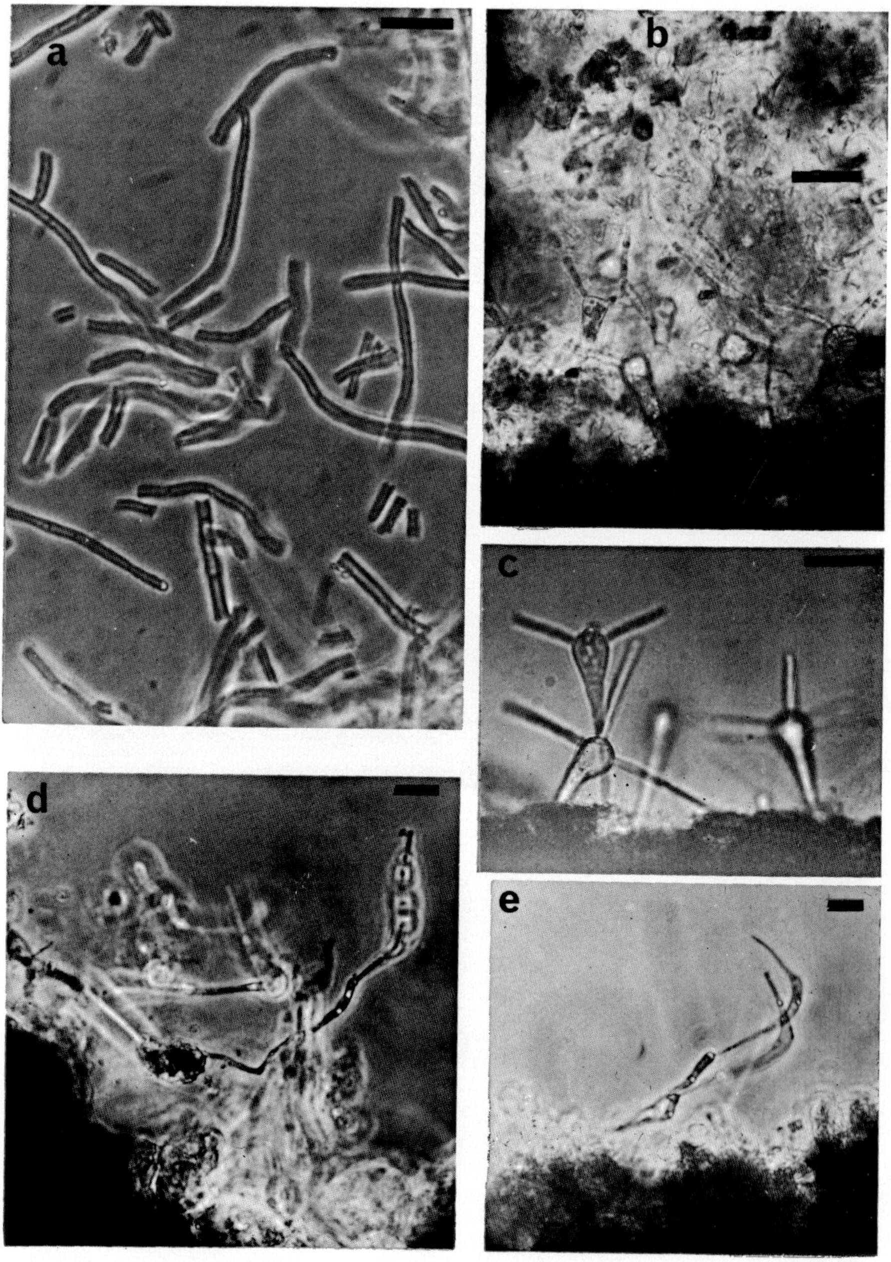

FIG. 5. Fungi. (a), (b) Squashed faecal pellets from *Gammarus pulex* following experimental feeds on *Tricladium giganteum* and *Clavariopsis aquatica* respectively; (b) undigested spores are apparent; (c) *Clavariopsis aquatica* spores as observed on a leaf collection; (d), (e) residual developing spores left on leaves following paddle agitation. The scale lines represent 30 μm.

sparse it may be advisable to reduce the incubation volume, necessitating the use of a Stomacher 80 for the subsequent paddle agitation. It is a disadvantage of the method that counting and identification are made under fairly unfavourable conditions of microscopy, where the developmental stages are not represented. However, examination of the paddled leaf material usually shows developing spores which are not detached (Fig. 5d, e) and this helps considerably. Detached mycelium is sometimes conspicuous in the counting area (Fig. 4b) and an estimation of its total linear extent, using methods such as those devised by Olson (1950), may be a possibility.

The attractiveness of a simple biochemical index for the presence of fungi in natural materials has been recognized for some time and the knowledge that the cell walls of fungi contain chitin while the cell walls of higher plants do not, is the current basis for investigation. As far as fungi which may be present in submerged leaves are concerned this generalization holds for the aquatic hyphomycetes but not for the biflagellate phycomycetes such as *Achlya*, *Lagenidium*, *Phytophthora*, *Pythium* and *Saprolegnia*, the walls of all of which are predominantly of cellulose (Aronson, 1965). It is true that *Phytophthora* and *Pythium* especially can often be isolated but the visual evidence is that the aquatic hyphomycete component is predominant and seemingly the most likely to figure in invertebrate diets. Therefore it remains of interest to make quantitative estimations of chitin in naturally decaying leaves and to use these to derive comparisons of fungal activity in material which has had different times of exposure, etc. Chitin is a long linear molecule, constituted entirely of β-1, 4 linked *N*-acetylglucosamine residues, and it may be hydrolysed completely, using strong mineral acid, to glucosamine. The latter is then estimated quantitatively (Swift, 1973). In another chemical method the chitin is partly deacetylated to chitosan in alkaline conditions and the glucosamine derived from this is converted to the aldehyde which is then determined quantitatively (Ride and Drysdale, 1972). Enzymic methods for chitin hydrolysis have a longer history and have tended to be eclipsed but their much greater simplicity in comparison with chemical methods makes them very attractive to the investigator who has limited time to expend on this facet of some wider ecological problem. Enzymic hydrolysis has the advantage that it detects *N*-acetylglucosamine whereas glucosamine detected by acid hydrolysis may have been derived from cell material other than chitin. With acid hydrolysis critical timing and control for the reaction is necessary to prevent complete destruction of the substrate. The enzymic method described below is based on that of Tracey (1955).

Puffballs (*Lycoperdon* spp.) which are well developed but have not yet produced spores are collected and placed in plastic bags. Internal deliquescence releases the brown coloured enzyme which accumulates in the bags over a week or ten days. This enzyme is then drained off and more is squeezed out of the fungi themselves. It is centrifuged at 4000 rev/m and further physical purification is made by filtering through a 0·8 μm membrane. This filtration is expected to remove all cell debris

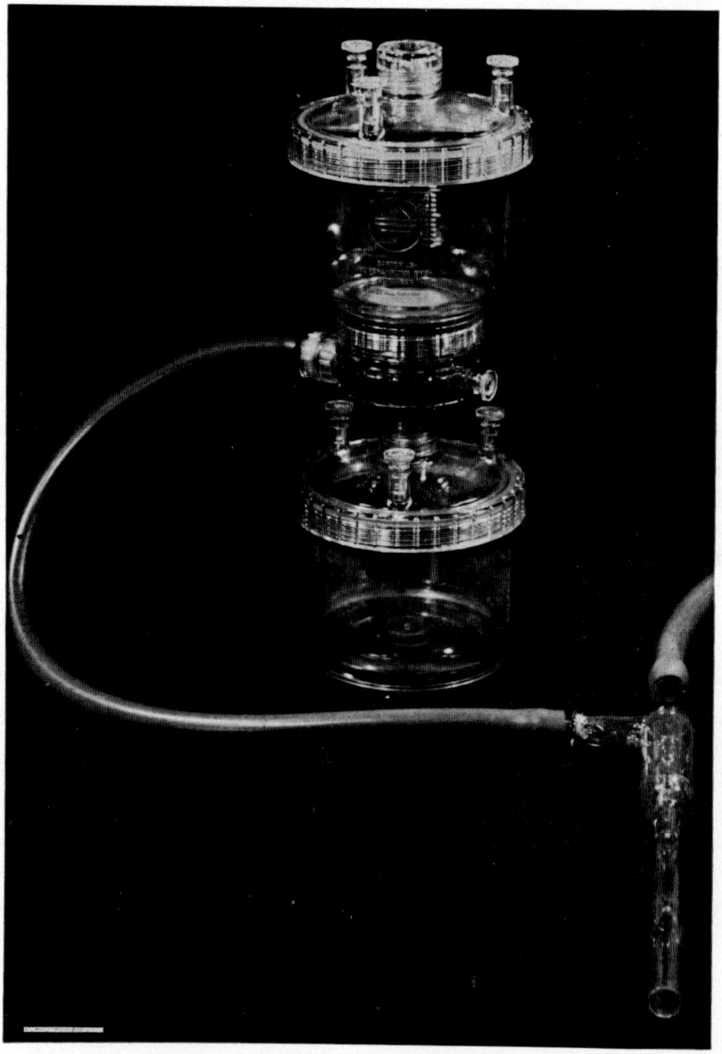

FIG. 6. Sartorius Steritest apparatus for membrane filtration. The scale line represents 3 cm.

and prevent any possible continued activity of the enzyme in storage. In this, as in other small-scale filtrations, the Sartorius Steritest apparatus (V. A. Howe and Co., Fig. 6) with standard 46 mm diam. Millipore membranes, has proved extremely useful. The enzyme is then dialysed in Visking tubing suspended in two litres of distilled water in a large glass dish. During the first 24 h the dialysate becomes brown in colour. This loss does not affect the subsequent performance of the enzyme but it is considered that a longer dialysis time with a water change may do so (Fig. 7). It therefore seems advisable to use the enzyme after only 24 h dialysis and accept the presence in it of some residual *N*-acetylglucos-

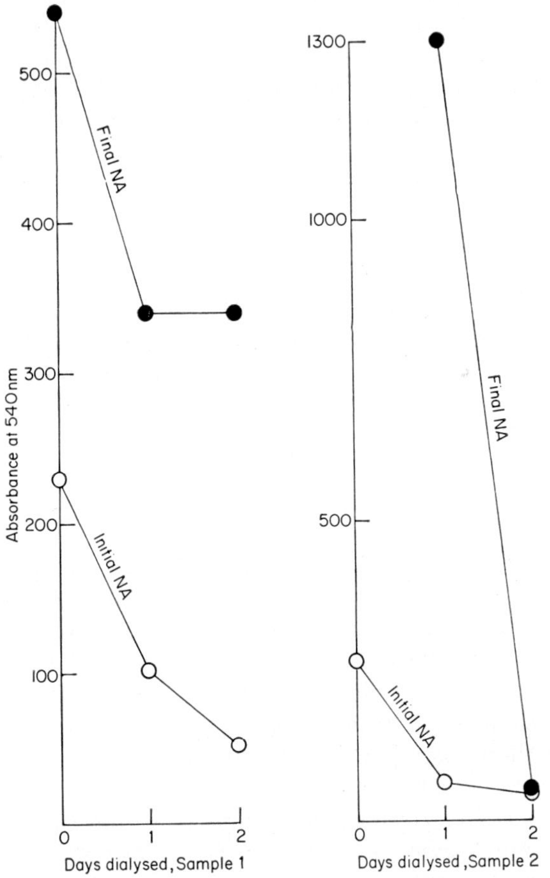

FIG. 7. The effect of dialysis on puffball chitinase activity. Initial NA relates to *N*-acetylglucosamine detected initially in the sample; Final NA relates to *N*-acetylglucosamine detected after eight days incubation in contact with a standard culture of fungus.

amine. Account must be taken of this by running the enzyme alone as a blank control when making the chitin estimations.

The optimal pH for chitinase activity is reported to be 4·1 (Ride and Drysdale, 1971) or 5·0 (Skujins *et al.*, 1965; Tracey, 1955). Buffering the reaction with a mixture of equal quantities of 0·08 M sodium acetate and 0·08 M acetic acid has proved effective. This solution (pH 4·8) and an equal quantity of the purified enzyme solution is placed on the leaf material in a stoppered vial and incubated at 33°. Test fungi and fungal-degraded leaves have been autoclaved in water before exposing them to the enzyme and any further microbial development during incubation is prevented by the addition of Thomersal (sodium merthiolate) at the concentration of 0·1 ml of a 0·1% solution per 2 ml of solution in the vial. According to Tracey (1955), enzyme solutions can be stored under toluene at 5°.

To make the *N*-acetylglucosamine determinations on standard solutions, enzyme controls and reacted fungal preparations, 2 g Ehrlich reagent (4-dimethylaminobenzaldehyde) is dissolved in 100 ml glacial acetic acid with 5 ml conc. HCl. Preliminary purification of the reagent, recommended by Tracey (1955), does not appear to be necessary. A standard solution giving 8 μg *N*-acetylglucosamine ml^{-1} is prepared and 1 ml is transferred to a 10 ml graduated centrifuge tube. Smaller aliquots of the standard solution, with water added, again to 1 ml, give 6 μg and 4 μg standards and 1 ml of the reacted preparation to be tested is included in the set together with a water blank. 0·3 ml of saturated sodium borate is added to each tube, and the tube end is covered with aluminium foil. The tubes are placed in a beaker of boiling water for 7 min and cooled. Glacial acetic acid is added to the 10 ml mark followed by 1 ml of the Ehrlich preparation. Tube contents are well mixed with a glass rod and after 45 min the absorbance at 540 μm is read on a Beckman 25 spectrophotometer with digital read-out. Testing of up to 100 μg of D-glucosamine hydrochloride has confirmed that there is no reaction with this molecule. It may be necessary to dilute the reacted preparation to place it on the standard curve and experience has shown that standard determinations vary quite widely in different solutions and even in different determinations made from the same solution (Fig. 8). It seems very possible that these discrepancies arise partly through errors in handling small volumes in standard 1 ml graduated pipettes and it might be advisable to use a 500 μl hypodermic syringe. However, since this technique is clearly of much more use for general comparative determinations rather than absolute determinations, such discrepancies need not be too worrying. From the results obtained it appears that enzyme activity can continue for at least 10 days, possibly much longer, and it is not certain that it is realistic to sample until a constant *N*-acetylglucosamine figure is attained. If the leaf material is stored after autoclaving in water, specimens with a different history of exposure in the field, etc., can be brought together to be reacted with one batch of the enzyme at the same time. Frequent monitoring allows a decision to be made on the duration of the reaction time.

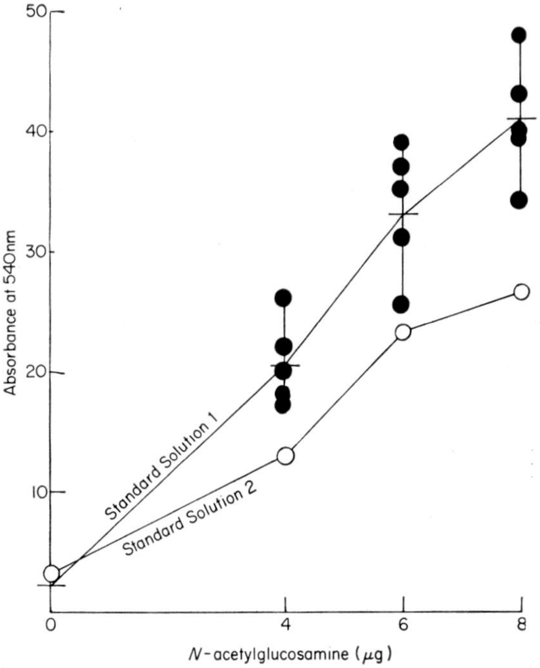

FIG. 8. *N*-acetylglucosamine determinations run as standards for comparative estimations of chitin in fungal-degraded leaves. For standard Solution 1 all the determinations and the means are plotted; for standard Solution 2 only the means are plotted.

Field baiting for fungi with artificial substrata

Since leaves and wood are largely cellulosic in nature it might be expected that the deposition of pure cellulose in streams would yield a representative sample of the leaf mycoflora. However, from experimental depositions of filter paper and cellophane, made in glass cylinders closed at both ends with fine muslin and enclosed in perforated polythene bottles, representatives of the Chytridiales and the Pythiaceae (Fig. 3c) were obtained by no aquatic hyphomycetes (Willoughby and Redhead, 1973). This implies that cellulase production is not a strongly marked characteristic of the aquatic hyphomycetes, an implication supported by evidence from cultural studies (Willoughby and Archer, 1973). However, in this type of deposition work it is important to retain an awareness that essential elements such as nitrogen, phosphorus and potassium must be extracted from the external environment, i.e. the water itself, if the mycoflora is to grow vigorously. It may be that the representatives of the Chytridiales and Pythiaceae are more efficient in such extractions than are the aquatic hyphomycetes.

Consumption of fungi by invertebrates

In a hypothetical natural situation where the microbial component on leaves is cropped by the invertebrates as soon as it develops, the animal intestines will be the sole repositories of information on the colonization pattern. Although such an extreme situation is unlikely to occur it remains of interest to gain information from this source. The consumption and digestion of small algae by leaf-eating invertebrates such as *Gammarus pulex* has been followed by isolating cells from faecal material and determining their viability. A low or nil viability is suggestive of cell contents having been removed and digested while a high viability implies that the cell has passed through the gut intact (Moore, 1975). Similar work has not been attempted with fungal diets although observations on faecal pellets produced by *G. pulex* which has been fed experimentally with *Tricladium giganteum* show chopped mycelium with no obvious cell contents (Fig. 5a). It is known that this fungus can sustain the animal indefinitely although it does not confer a high growth rate (unpublished data). With *Clavariopsis aquatica* as food, undigested spores are seen in the faeces (Fig. 5b). *Gammarus pulex* and many other invertebrates produce faeces in discrete packets and an animal collected from nature can be quarantined for several days without food, during which time its bodily functions will continue. Faecal pellets can be removed and examined with the prospect that microscopial examination or viability tests might yield information on the previous microbial feeding regime.

Fungi already present at leaf fall

The possible significance of fungi which are already active on the leaf before it falls into water cannot be discounted. According to Lubyanov and Zubchenko (1970), *Gammarus balcanicus* feeds preferentially on black-spot leaf areas of sycamore and maple which are infected and partially degraded by *Rhytisma acerinum*. Little is known of the prospects of this and other similar parasitic or phylloplane fungi to continue their activity in water. According to Newton (1971) the common phylloplane fungus *Aureobasidium pullulans* disappeared within a month from alder, elm, oak and willow leaves deposited in the River Lune, but personal experience of the very frequent isolation of this particular fungus from leaves in varying stages of decay suggests that it may have an active aquatic role. Background knowledge of the sub-aerial fungal component of leaves is readily obtained by direct microscopy and by plating leaf washes or macerations on to GYS–penicillin, streptomycin agar (see below).

Actinomycetes

Traditionally the Actinomycetales have been regarded as a group having little or no affinity with freshwater; however, this conception is largely biased by the knowledge that the Streptomycete component is dry spored. Sporangial forms releasing spores which are motile in water, and placed in the Actinoplanaceae, are readily obtained from littoral leaf litter by damp incubation of the material (Fig. 9a–c) or by enrichment baiting with pollen. It was later shown that these forms could also be obtained by washing the freshly collected leaves in sterile water and plating the washings on to chitin-actidione agar (Fig. 2b). Enrichment systems were also productive; when leaf debris was set up in shallow dishes in the laboratory, Actinoplanaceae could often be recovered from agar platings made from the surface water over several days. Other forms obtained from the agar plating of leaf washes have motile spores borne directly on phialides, and critical sectioning and microscopy is necessary to resolve those which grow below the agar surface (Fig. 9d) (Willoughby, 1969a). These phialidic forms are apparently previously undescribed (see Cross and Goodfellow, 1973) and as numerous soil studies have failed to elicit them, the strong implication arises that aquatic decaying leaves constitute their true ecological niche. In addition to the actinomycetes which produce sporangia or spores directly on the agar, and hence can be identified there, numerous sterile colonies are obtained from the agar plating of leaves. Many are pigmented, suggesting that they are putative Actinoplanaceae. Laboratory manipulations to induce sporulation include transferring portions of colonies to water, subculturing on to media containing humic acids (Willoughby *et al.*, 1968) or cultivating the isolate with pollen floating on 1% starch solution (Kuznetzov, 1969).

Wood also apparently sustains the growth of aquatic actinomycetes, particularly *Actinoplanes*, and twig exposures to obtain them from streams are made in the same way as for fungi. The initial laboratory inoculations of the isolates obtained on to sterilized twigs in flasks generally failed, and this was ascribed to leaching of inhibitory tannin-like materials into the flask liquid. However, when twigs stripped of the bark and cortex

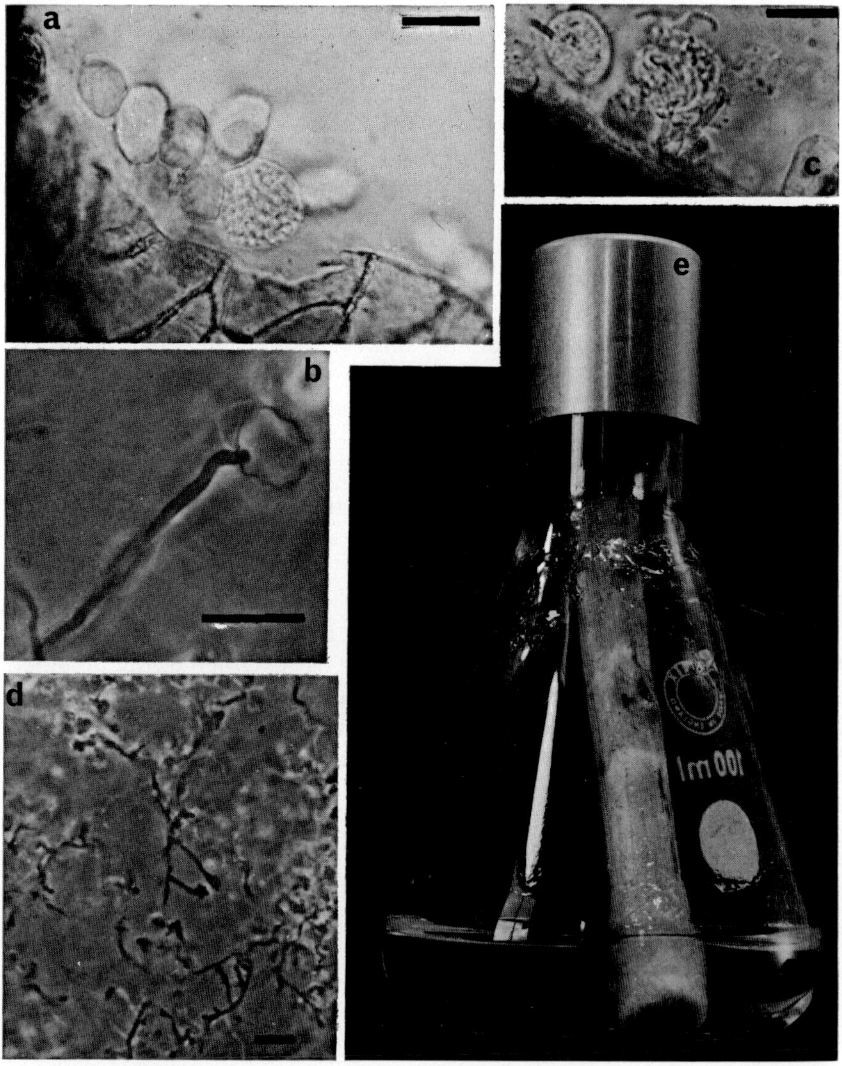

FIG. 9. Actinomycetes. (a), (c) Damp incubation of oak leaf material from Windermere littoral yielding *Actinoplanes* and *Spirillospora* respectively; (b) *Actinoplanes* sporangium dehisced; (d) phialidic form isolated from leaves and growing in chitin-actidione agar. The potentially motile phialospores are seen in small terminal clusters; (e) actinomycete growth in subculture on a twig with bark and cortex removed. In (a)–(d) the scale lines represent 10 μm.

were used there was often good growth on the wood, particularly if the flask liquid was largely devoid of leached sugars, detected by anthrone reagent (Fig. 9e). This system made it possible to study sporulation phenomena in recalcitrant isolates which had been hitherto sterile (Willoughby, 1971).

Actinophage

It might perhaps have been expected that specific virulent phage systems would accompany the actinomycetes in decaying leaves. This expectation is realized as far as *Actinoplanes* is concerned and *Actinoplanes* phage is readily obtained from littoral leaf litter (Willoughby *et al.*, 1972). On the other hand, benthic lake mud yields neither *Actinoplanes* nor its phage. If it is then accepted, provisionally, that host and phage activity is complementary in aquatic niches it can be seen that detection of phage in a particular situation may allow a prediction that the host is growing actively and is not merely present in an inert condition. Such a prediction is made for *Actinoplanes* in decaying littoral leaf litter. Conversely, failure to detect the phage of a particular taxon, in a situation where the host is present and viable, may allow a prediction that no active growth of that host is occurring. Such a prediction might have been made for nocardioform-Lspi, originally isolated readily from lotic and lacustrine situations and thought to be a truly aquatic form although its phage was never recovered (Willoughby, 1969b). It now appears that such a prediction would have been justified as has become apparent that nocardioform-Lspi is a wash-in form derived from the dung of herbivorous animals. Furthermore its phage is also readily isolated from the dung (Cross and Rowbotham, 1974). A study of the distribution of the phages of the phialidic types of actinomycetes isolated from leaves would be especially interesting, if it could be made, but unfortunately these hosts are extremely difficult to isolate and handle at present.

Bacteria

In addition to fungi and actinomycetes, allochthonous materials in freshwater also harbour bacteria. These bacteria may colonize the autolysing fungi (Fig. 10c) as well as the plant materials themselves. Filamentous bacteria are usually very conspicuous on decaying leaves, whether they be of trees (Fig. 10b) or of herbaceous plants (Fig. 10e), but the only knowledge of them appears to be through direct microscopy. It is believed that they have never been cultured nor recognized as an ecological group, although they may well parallel the aquatic hyphomycete fungi in

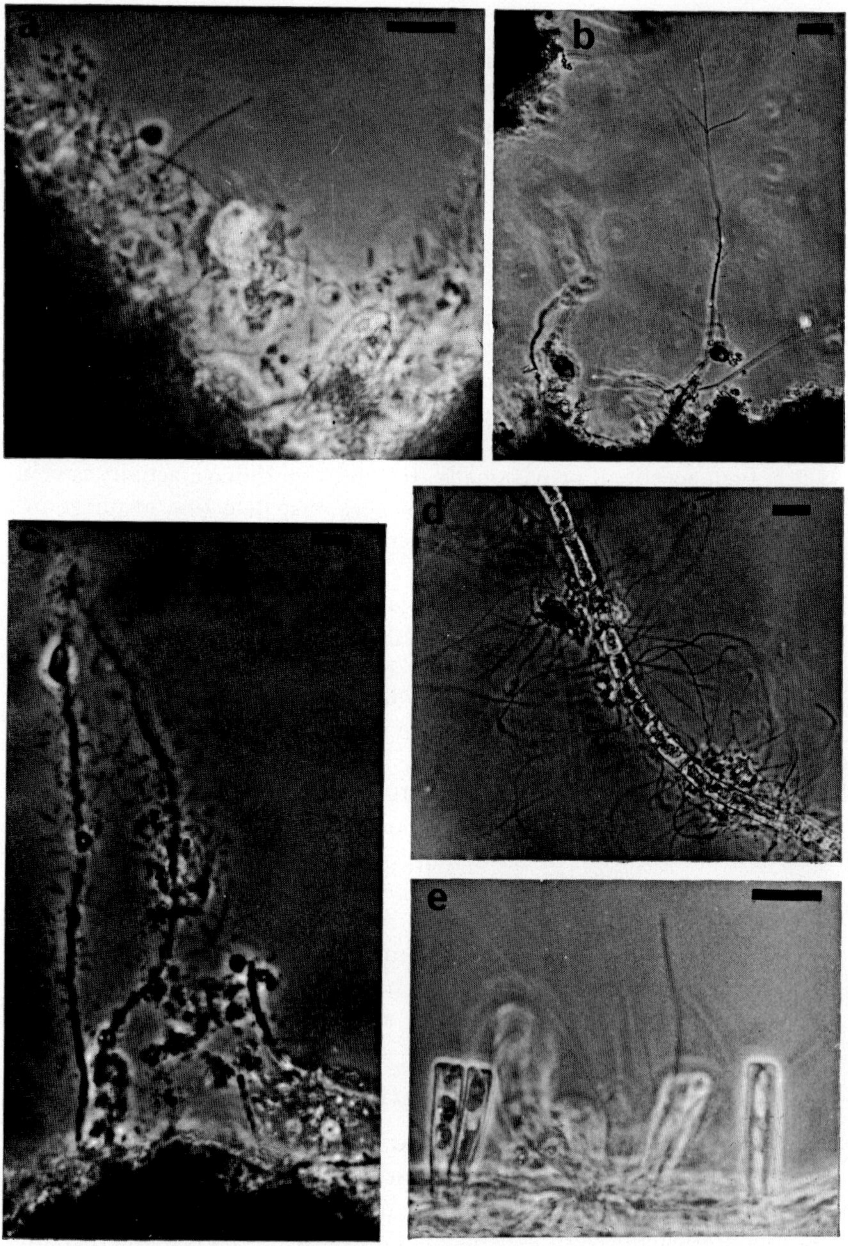

FIG. 10. Bacteria. (a), (c) Sycamore leaves in experimental continuous flow systems showing build up of (a), filamentous bacteria and (c), bacteria on autolysing fungal hyphae; (b) leaf of smooth leaved elm (*Ulmus carpinifolia*) collected from Smooth Beck, Lancashire, and showing growth of filamentous bacteria; (d) filamentous bacteria on the alga *Zygogonium ericetorum*; (e) filamentous bacteria and diatoms on grass leaves collected from Mosedale Beck, River Duddon. The scale lines represent 20 μm.

this respect. Experimental continuous flow systems using materials from nature often enhance their growth (Fig. 10a) and this is suggestive of adaptation to a flowing water regime. As with *Sphaerotilus natans*, the attached filamentous form may be highly efficient in nutrient abstraction under such conditions. Similar filamentous growths may also appear on stream algae (Fig. 10d). Here they have been noticed most particularly on algae kept in the laboratory under conditions presumed to be unfavourable (poor illumination, etc.) and their appearance seems to result from artificial "enrichment".

Growth Media

GYS–penicillin, streptomycin agar for fungi is a very useful growth medium for general isolation, including Saprolegniaceae etc. (g $litre^{-1}$): glucose 10, soluble starch 5, yeast extract 2, $Na_2HPO_4.12H_2O$ 0·6, KH_2PO_4 2·04. Make up in flasks together with weighed agar addition (2%). Autoclave and remove immediately pressure is down to add the weighed antibiotics, sodium benzylpenicillin and streptomycin sulphate (Glaxo) at a concentration of 0·05 g of each 100 ml^{-1}. Cool and pour. Add antibiotics similarly for malt (1%) penicillin, streptomycin agar.

Acknowledgements

My thanks are due to Dr. P. A. Cranwell and Mr. E. Rigg for chemical advice; to colleagues and friends, particularly Miss Mary Suart, for enthusiastic puffball collection; to Prof. G. W. Minshall for introducing me to the tube exposure of leaves.

References

Aronson, J. M. (1965). The cell wall. In *The fungi, an advanced treatise*, Vol. I, 49 (Ainsworth, G. C. & Sussman, A. S., eds). New York and London: Academic Press.

Bärlocher, F. & Kendrick, B. (1975). Assimilation efficiency of *Gammarus pseudolimnaeus* (Amphipoda) feeding on fungal mycelium or autumn-shed leaves. *Oikos*, **26**, 55.

Cross, T. & Goodfellow, M. (1973). Taxonomy and classification of the Actinomycetes. In *Actinomycetales: characteristics and practical importance* Sykes, G. & Skinner, F. A., eds). Soc. appl. Bact. Symp. Ser. No. 2. London and New York: Academic Press, p. 11.

Cross, T. & Rowbotham, T. J. (1974). The isolation, enumeration and identification of Nocardioform bacteria in clean and polluted streams, and in lake waters and mud. *Proc. 1st int. conf. on biology of the Nocardiae*. Venezuela: Merida, p. 48.

INGOLD, C. T. (1975). *An illustrated guide to aquatic and water-borne Hyphomycetes.* Scientific Publication 30, Freshwater Biological Association.

IQBAL, S. H. & WEBSTER, J. (1973). Aquatic Hyphomycete spora of the River Exe and its tributaries. *Trans. Br. mycol. Soc.*, **61**, 331.

JOHNSTON, D. W. (1972). *Actinomycetes in aquatic habitats.* Ph.D. Thesis, Univ. of Bradford.

KUZNETZOV, V. D. (1969). Cultivation of members of the family Actinoplanaceae on plant pollen. *Mikrobiologiya*, **38**, 143.

LUBYANOV, I. P. & ZUBCHENKO, I. A. (1970). Fundamental aspects of the feeding of the amphipod *Gammarus* (*R.*) *balcanicus* (Crustacea, Amphipoda). *Nauchn. Dokl. vyssh. Shkoly* (*biol. Nauk*), **7**, 12.

MOORE, J. W. (1975). The role of algae in the diet of *Asellus aquaticus L.* and *Gammarus pulex L. J. Anim. Ecol.*, **44**, 719.

NEWTON, J. A. (1971). *A mycological study of decay in the leaves of deciduous trees on the bed of a river.* Ph.D. Thesis, Univ. of Salford.

OLSON, F. C. W. (1950). Quantitative estimates of filamentous algae. *Trans. Am. microsc. Soc.*, **69**, 272.

RIDE, J. P. & DRYSDALE, R. B. (1971). A chemical method for estimating *Fusarium oxysporum f. lycopersici* in infected tomato plants. *Physiol. Pl. Path.*, **1**, 409.

RIDE, J. P. & DRYSDALE, R. B. (1972). A rapid method for the chemical estimation of filamentous fungi in plant tissue. *Physiol. Pl. Path.*, **2**, 7.

SKUJINS, J. J., POTGIETER, H. J. & ALEXANDER, M. (1965). Dissolution of fungal cell walls by a Streptomycete chitinase and β–(1→3) glucanase. *Arch. Biochem. Biophys.*, **111**, 358.

SWIFT, M. J. (1973). The estimation of mycelial biomass by determination of the hexosamine content of wood tissue decayed by fungi. *Soil Biol. Biochem.*, **5**, 321.

TRACEY, M. V. (1955). Chitin. In *Modern methods of plant analysis*, Vol. II (Paech, K. & Tracey, M. V., eds). Berlin: Springer-Verlag, p. 264.

WILLOUGHBY, L. G. (1966). An unusual chytrid from incubated leaf litter. *Trans. Br. mycol. Soc.*, **49**, 451.

WILLOUGHBY, L. G. (1969a). A study on aquatic Actinomycetes: the allochthonous leaf component. *Nova Hedwigia*, **18**, 45.

WILLOUGHBY, L. G. (1969b). A study of the aquatic actinomycetes of Blelham Tarn. *Hydrobiologia*, **34**, 465.

WILLOUGHBY, L. G. (1971). Observations on some aquatic Actinomycetes of streams and rivers. *Freshwat. Biol.*, **1**, 23.

WILLOUGHBY, L. G. & ARCHER, J. F. (1973). The fungal spora of a freshwater stream and its colonization pattern on wood. *Freshwat. Biol.*, **3**, 219.

WILLOUGHBY, L. G. & MINSHALL, G. W. (1975). Further observations on *Tricladium giganteum. Trans. Br. mycol. Soc.*, **65**, 77.

WILLOUGHBY, L. G. & REDHEAD, K. (1973). Observations on the utilization of soluble nitrogen by aquatic fungi in nature. *Trans. Br. mycol. Soc.*, **60**, 598.

WILLOUGHBY, L. G., BAKER, C. D. & FOSTER, S. E. (1968). Sporangium formation in the Actinoplanaceae induced by humic acids. *Experientia*, **24**, 730.

WILLOUGHBY, L. G., SMITH, S. M. & BRADSHAW, R. M. (1972). Actinomycete virus in fresh water. *Freshwat. Biol.*, **2**, 19.

Sewage Pollution and Shellfish

P. A. AYRES, H. W. BURTON AND MARY L. CULLUM

Ministry of Agriculture, Fisheries and Food, Fisheries Laboratory, Burnham-on-Crouch, Essex, England

Introduction

The disposal of domestic sewage to estuarine and coastal areas may lead to contamination of commercially important shellfish which are found there.

Particularly at risk are bivalve molluscs which, because of their mode of feeding, may accumulate bacteria and viruses of faecal origin, often to far in excess of the levels found in the surrounding water. Organisms of faecal origin may be reduced by sewage treatment and by a variety of factors, once released into the marine environment but large numbers of bacteria, and perhaps viruses, may still persist to be ingested by molluscs. These molluscs may be subsequently harvested for human consumption, often as raw or partially processed products, and so transmit bacterial and viral infections direct to man.

In trying to make an assessment of existing and potential public health hazards posed by the consumption of shellfish taken from polluted waters, a broad understanding of the bacteriology of shellfish from polluted and unpolluted sites is necessary. The paper presents a brief review of the problems involved and of some of the research undertaken to investigate aspects of sewage pollution and shellfish.

The Nature of the Problem

In many estuarine and coastal areas it has long been the practice to dispose of domestic sewage effluent direct to the sea. Biological effects such as reduction or cessation of growth and reproductive ability or in extreme cases, survival, of some components of plant and animal life may be evident at the site of discharge. Because of the dilution which an effluent subsequently receives such obviously detrimental effects are rarely apparent elsewhere and the type of treatment the effluent receives assumes greater importance. Primary treatment such as screening, maceration and sedimentation may remove gross solids and suspended material which might constitute a visual nuisance from an amenity or recreational viewpoint. Secondary treatment further to reduce suspended solids and to change the chemical nature of the effluent is widely applied, but neither treatment will have a significant effect in reducing the bacterial content. Tertiary treatment methods must be applied when high bacteriological quality is an important criteria but with the exception of chlorination no method will produce a bacteria-free effluent. Though chlorine is effective at removing the bacteria, release of chlorinated effluent may, in itself, create an environmental problem in terms of its toxicity to plant and animal life, particularly larval and juvenile stages.

Therefore, even with high standards of sewage treatment, large numbers of micro-organisms including bacteria, protozoa and viruses may be discharged into areas used for shellfish cultivation. These micro-organisms are generally of little or no public health significance, but may include pathogens or organisms potentially capable of producing illness or disease in humans. Of particular importance are bacteria of the Salmonella group (especially *S. typhi* and *S. paratyphi*), *Shigella* spp. and *Vibrio cholera*, all of which have been associated with shellfish transmitted disease in the United Kingdom and/or Europe. Enteroviruses and the agent of infectious hepatitis may also present a potential hazard.

Most of the major molluscan shellfisheries of the United Kingdom are centred in estuarine and shallow coastal areas where hydrographic conditions are suitable for their growth and it is this coincidence with suitable sites for effluent discharge which may constitute a public health problem. In broad terms, commercial molluscan shellfish may be divided into two main groups, the bivalves or filter feeders, with two shell valves, and the gastropods, having a single snail-like shell. Each group may acquire contamination from sewage but because of differences in the mode of feeding it is the bivalves which represent the major public health problem.

The Role of Molluscan Shellfish

Two species of gastropod are exploited commercially in England and Wales, the whelk (*Buccinum undatum*) and the periwinkle (*Littorina littorea*). The whelk is carnivorous, feeding by means of an extensible mouth or proboscis, and the winkle, a herbivore, feeds by rasping or grazing plant material from rocks and stones. Both animals may acquire low levels of faecal contamination by eating contaminated food or by trapping polluted water in the shell cavity. Reynolds (1957) has shown that coliform bacteria in winkles will be destroyed by immersing the animals in boiling water for 30 s; at least 1½ min is necessary to make the flesh extractable from the shell and available for consumption. The whelk, a larger animal, requires 8 to 10 min boiling to permit extraction of the meats but coliform bacteria are killed in under 3 min (Ayres, unpublished). Therefore, although it may be possible to obtain polluted gastropods these do not normally constitute a public health problem if properly prepared for consumption.

In terms of commercial importance and gross value, the bivalve shellfish form the major group of commercial molluscs and those species exploited include three species of oyster, mussels, cockles, clams and escallops. Some may be regarded as exclusively estuarine in habit but the escallop, for example, is more common in offshore coastal waters. In spite of differences in gross morphology or appearance, all bivalves have a common mode of feeding, by filtration. A diagrammatic view of the European flat oyster (*Ostrea edulis*) is shown in Fig. 1 in order to illustrate the body plan of a typical bivalve filter feeding mollusc.

There are four layers of gills, one next to another; each layer is composed of complex folds of tissue richly supplied with blood vessels

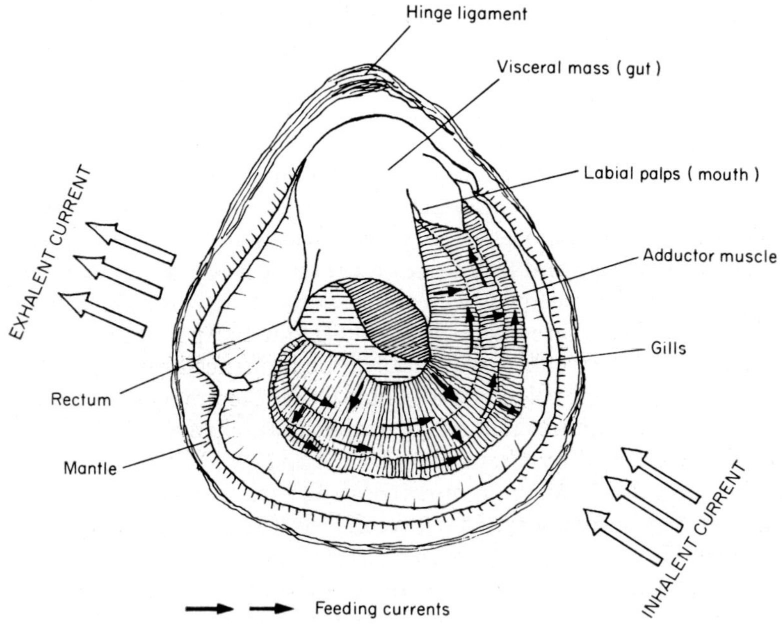

FIG. 1. Diagram of the oyster (*O. edulis*) to illustrate general body plan and mode of feeding (flat shell valve and mantle removed).

and covered in cilia. The cilia beat with a rhythmic wave-like motion and induce a current of water to flow into the shell cavity and over the gills. Oxygen required for respiration is removed at the water/tissue interface while the cilia trap any small particles of suspended matter. This suspended material consists of silt and organic detritus together with living material such as phytoplankton and bacteria. Bound with mucus, this material passes to the gill margins and along to the mouth for ingestion. Some coarse material is rejected, but the remainder is passed to the gut and any digested or waste material expelled through the rectum and carried clear of the animal by the exhalent water current. As a mollusc such as the oyster can filter up to five litres of water an hour under optimal conditions, large quantities of particulate matter can be ingested. In a situation where the surrounding water is polluted by sewage, bivalves may ingest and concentrate faecal bacteria (including potential pathogens), viruses and other micro-organisms. Studies with the European flat oyster (*O. edulis*) indicated a concentration factor ten times the levels in surrounding water at 16° (Wood, 1965).

Factors Affecting Sewage Pollution of Bivalve Molluscs

Once an effluent is released from an outfall, its bacterial and viral content is reduced by a complex array of physical, chemical and biological factors. Much of the available data concerns the fate of coliform bacteria, or more specifically *Escherichia coli*, as indicators of what may occur with pathogenic bacteria. Pathogens if present, are likely to occur in relatively low numbers compared with coliforms and therefore pose certain problems when the objective is to enumerate them in the environment. Problems of this kind are even more obvious with viruses as these are also present in very low numbers and techniques for their recovery are more complex.

Dilution by the receiving water may be sufficient to reduce bacterial levels to a point where shellfish remote from the pollution source are unaffected. Dilution effects will, in turn, be influenced by factors such as water depth, current velocity and wind. Other factors which may be important in affecting actual survival, rather than changes in distribution, have been reviewed by Carlucci and Pramer (1959). These include adsorption and sedimentation, nutrients, natural antibiotics, phage and predators, pH and salinity. More recent work by Gameson *et al.* (1968) and Reynolds (1965) has also demonstrated the importance of sunlight. Removal of bacteria by filter feeding animals may be locally important, but the significance of the process lies in the public health hazards resulting from the uptake of potentially pathogenic micro-organisms by bivalve molluscs which may be taken for human consumption. Given that the

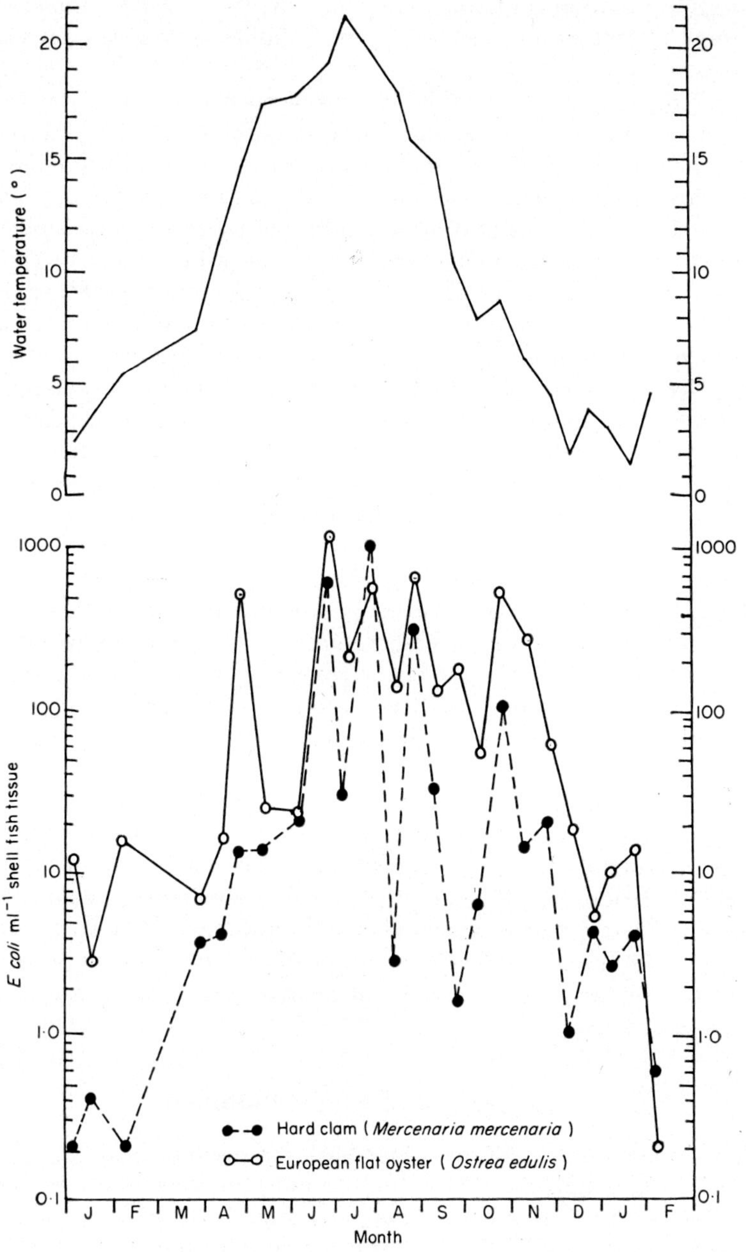

FIG. 2. *E. coli* levels in bivalve molluscs sampled at intervals from a polluted area.

nature and extent of contamination which shellfish receive is governed by a variety of factors outlined above, the animals themselves are equally important.

It has already been noted that the gills of bivalve molluscs serve a dual respiratory and feeding role and filtration is therefore closely linked to metabolic activity of the animal. This in turn is temperature related, so that at low temperatures metabolism and respiration proceeds almost at maintenance level and filtration and bacterial uptake is correspondingly reduced. In consequence, shellfish held in an area subject to constant levels of water borne sewage pollution will be more highly polluted during summer than in winter. Figure 2 illustrates some field observations of *E. coli* levels in the European flat oyster (*O. edulus*) and the hard clam (*Mercenaria mercenaria*) taken at intervals throughout the year from a polluted area. Regression analysis of these data gives a significant correlation ($p = 0{\cdot}001$) between *E. coli* levels in the shellfish and water temperature. Various species of shellfish have different temperature requirements and the figure shows that hard clams with a higher optimal temperature requirement are much less active during winter months and therefore acquire fewer faecal bacteria. Of the commercially exploited shellfish, the mussel (*Mytilus edulis*) is probably the most hardy and will be active even at a water temperature of 0°. While temperature is singularly the most important factor governing activity of bivalve shellfish, the species concerned also have varying requirements for salinity and dissolved oxygen and these will also exert an influence on uptake of contamination via filtration.

In practical terms, a useful technique for assessing the influence of environmental factors on shellfish activity is to measure the filtration rate of the animals using rate of removal of neutral red dye as an index (Cole and Hepper, 1954). The purification of shellfish to remove faecal bacteria (Wood 1969) is essentially a continuation of normal feeding activity in the absence of faecal pollution and so an understanding of the parameters influencing bacterial uptake also has applications in consideration of bacterial removal.

Assessment of Faecal Pollution

Although there are no statutory standards for sanitary quality of bivalve shellfish in England and Wales, the recommendations of Sherwood and Scott Thomson (1953) are widely applied. Standards in use are shown in Table 1, together with the equivalent result by the percentage clean method of Knott (1951) and recommended action. With the advent and widespread use of purification systems, it is expected that treated shellfish

TABLE 1. Bacteriological standards for molluscan shellfish in England and Wales

E. coli g^{-1} tissue	Equivalent percentage clean (Fishmongers' Co.; Knott, 1951)	Grade (Sherwood and Scott Thompson, 1953)	Action taken
0–2	100		
3–4	90	I	Sale permitted
5	80		
6–15	70	II	Temporary prohibition pending further samples
16–25	60	III	Sale prohibited
>26	50		

would contain less than 2 *E. coli* g^{-1} tissue. Further, as a result of the improvement in laboratory testing procedures (Ayres, 1975a) there has been a significant improvement in quality without adjustment of the standard.

As molluscan shellfish have the ability to concentrate bacteria they are useful tools in surveying areas for evidence of faecal pollution and represent an extra dimension in sampling which often cannot be achieved by bacteriological sampling of water alone. Surveys of estuarine areas can yield information which may be applied in two ways. First, the determination of sources and zones of influence of existing sewage discharges and the most advantageous site of proposed discharges. Secondly, the sanitary assessment of an area of existing or potential shellfish culture to determine whether an area warrants control under the Public Health (Shellfish) Regulations 1934. Figure 3 represents the mean results of

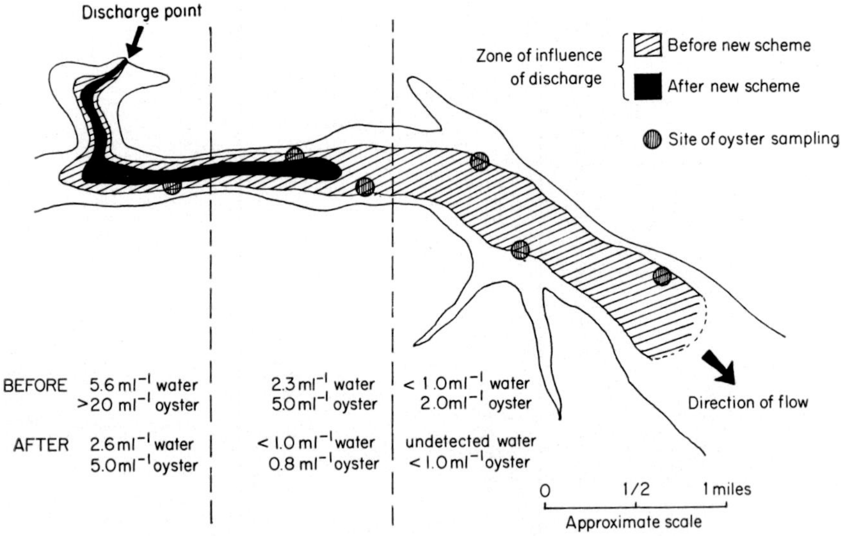

FIG. 3. Mean results of oyster/water surveys carried out in an area before and after introduction of a sewage works improvement scheme.

oyster/water surveys carried out in an area before and after introduction of a sewage works improvement scheme. Samples were taken at intervals over a six-month period to minimize seasonal effects and also at or near low water when pollution levels were judged to be maximal. In this instance, it was possible to show significant improvements in water and shellfish quality as the result of sewage plant modernization and resiting of the discharge point. Methods employed for such surveys include a membrane filtration technique for waters (Halls and Ayres, 1974) and the modified MacConkey roll-tube technique (Reynolds and Wood, 1956) using the machine maceration method (Ayres, 1975a) for shellfish.

In addition to *E. coli*, the coliform group generally, *Clostridium welchii* and the faecal streptococci or enterococci have been proposed, and used, as indices of faecal pollution. Investigations in this laboratory suggest that although tests for these other organisms may be usefully applied in specific situations, their value is to complement, rather than replace, the use of *E. coli* (Ayres, 1975b).

Shellfish Bacteriology Baseline Studies

Although *E. coli* is used as an indicator of faecal contamination, sewage may be expected to contribute large numbers of other organisms to the marine environment and these may also be ingested by molluscan shellfish. The possible significance of this input for shellfish bacteriology has been the subject of continuing research in this laboratory prompted by the need to understand what constitutes the "normal" bacterial flora of shellfish. Apart from transmission of identifiable diseases such as typhoid, cholera and infectious hepatitis, shellfish are occasionally associated with gastroenteritis of unknown aetiology (Preston, 1968; Gunn and Rowlands, 1969), which may be linked with sewage pollution (Ayres, 1971).

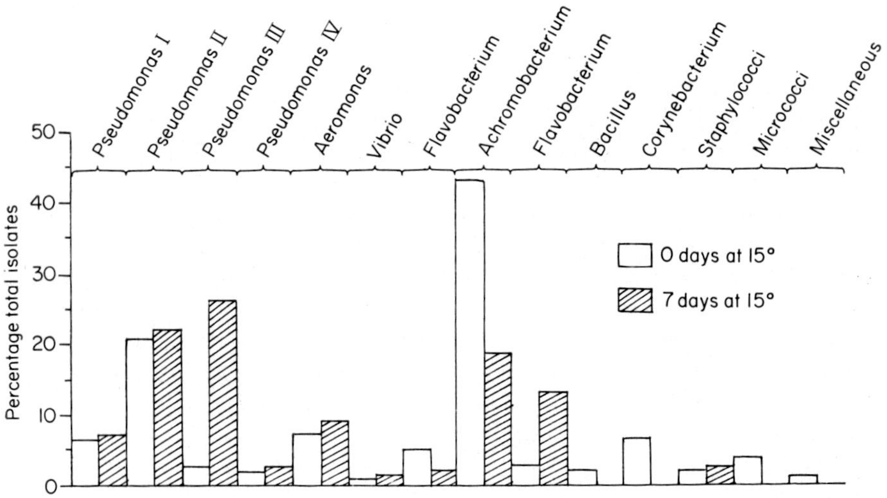

FIG. 4. The effect of air storage on the bacterial flora of oysters.

Some limited qualitative work has been undertaken, including the effects of air storage on the bacterial flora of oysters. Results from initial investigations are presented in Fig. 4, suggesting that marked changes occur in the dominant bacterial flora as judged by application of the identification scheme proposed by Shewan (1963). Although such studies indicate important gross changes in the bacterial flora, taxonomic problems in classifying the diversity of organisms encountered limit the value of such investigations. In consequence, much of the work has been quantitative in nature and for this purpose nutrient agar and Zobells 2216 medium have been adopted for total plate counts (TPC) of shellfish and water respectively.

Total plate count estimates of bacteria in oysters and water from a site in the River Crouch, Essex over a period of two years indicated that the numbers in water ranged from 10^1 to 10^5 ml^{-1} and in oysters from 10^2 to 10^6 g^{-1}. The concentration ratio of oyster to water counts gave a highly significant correlation with water temperature ($p = 0{\cdot}001$) confirming earlier results with *E. coli* (shown in Fig. 2) and indicating that under natural conditions bacterial count is primarily a function of shellfish activity. The fact that clean and fresh oysters may already have TPC exceeding US standards of 5×10^5 g^{-1} for marketable shellfish prior to harvesting, led to a study of the bacteriological quality of shellfish sampled at market level in the United Kingdom (Ayres, 1975b). This demonstrated that some 50% of the samples taken, although acceptable by application of the *E. coli* standards shown in Table 1, would fail the standard plate count criterion used in the USA and suggests that purely quantitative standards are of little value without epidemiological or qualitative support.

Further quantitative studies have been undertaken to explore the hypothesis that oysters and other molluscs have a "commensal" or basic bacterial population which cannot be reduced below levels characteristic of the areas from which the shellfish are harvested, i.e. a bacterial population typical of the animal rather than its environment, but which varies from area to area. The bacterial densities observed may correspond with the baseline or winter levels observed in the seasonal studies referred to earlier since both are of the same order. Studies of quantitative changes using oysters subject to purification procedures, storage in air and in water reveal broad population trends similar to those shown in Fig. 4. The bacterial content of oysters tends to equilibrate to a level corresponding to 10^4–10^5 g^{-1} regardless of treatment or initial TPC. Such changes occur more rapidly at 20° than at 10°, and at the higher temperature will then increase as spoilage ensues.

Preliminary observations in which oysters are held in water with sewage to simulate natural exposure to a polluted area have been made. It was anticipated that increases in TPC particularly at 37° would be evident, and although the animals obviously accumulated *E. coli*, no increases in TPC were observed; occasionally TPC actually decreased. A closer examination showed that oyster activity was stimulated by the presence of organic material in the sewage, resulting in removal of bacteria from the oyster gut at a rate exceeding that of bacterial uptake. There are obviously a number of closely related factors governing the bacterial content of shellfish and more research is necessary to elucidate the nature of these factors and the inter-relationships involved.

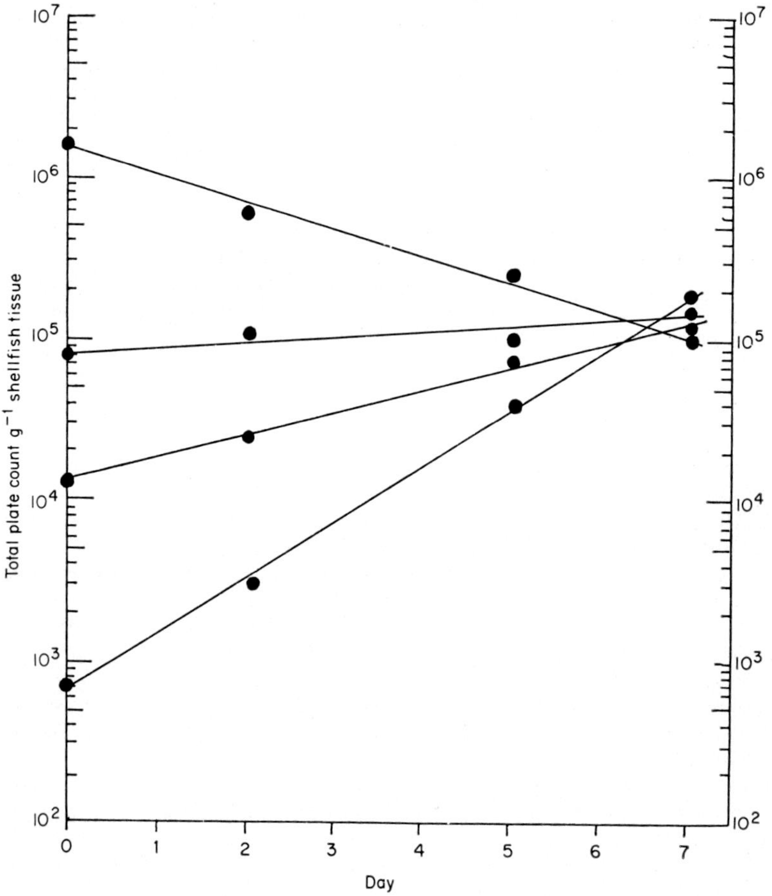

FIG. 5. The influence of initial bacterial count on subsequent changes during storage of oysters in water at 20° for seven days.

References

AYRES, P. A. (1971). *Non-specific illness associated with the consumption of molluscan shellfish: a report of current investigations in England.* International Council for Exploration of the Sea, CM 1971, Shellfish and Benthos Committee, Doc. K:17 (mimeo).

AYRES, P. A. (1975a). Recovery of *Escherichia coli* and coliforms from macerated shellfish. *J. appl. Bact.*, **39**, 353.

AYRES, P. A. (1975b). The quantitative bacteriology of some commercial bivalve shellfish entering British markets. *J. Hyg., Camb.*, **74**, 431.

CARLUCCI, A. F. & PRAMER, D. (1959). Factors affecting the survival of bacteria in seawater. *Appl. Microbiol.*, **7**, 388.

COLE, H. A. & HEPPER, B. T. (1954). The use of neutral red solutions for the comparative study of filtration rates in lamellibranchs. *J. du Cons. L'Explor. Mer.*, **20**, 197.

GAMESON, A. L. H., PIKE, E. B. & BARRETT, M. J. (1968). Some factors influencing the concentration of coliform bacteria on beaches. *Revue int. Oceanogr. med.*, **9**, 255.

GUNN, A. D. G. & ROWLANDS, D. F. (1969). A confined outbreak of food poisoning which an epidemiological exercise attributed to oysters. *Medical Officer*, **122**, 75.

HALLS, S. & AYRES, P. A. (1974). A membrane filtration technique for the enumeration of *Escherichia coli* in seawater. *J. appl. Bact.*, **37**, 105.

KNOTT, F. A. (1951). *Memorandum on the principles and standards employed by the Worshipful Company of Fishmongers in the bacteriological control of shellfish in the London markets.* London: Fishmongers' Co.

PRESTON, F. S. (1968). An outbreak of gastroenteritis in aircrew. *Aerospace Medicine*, **39**, 519.

REYNOLDS, N. (1957). The effect of heat on the *Bact. coli* content of periwinkles. *Mon. Bull. Minist. Hlth*, **16**, 86.

REYNOLDS, N. (1965). The effect of light on the mortality of *E. coli* in seawater. *Pollutions marines par les micro-organismes et les produits petroliers*, Symposium de Monaco (avril 1964). Monaco: Commission Internationale pour l'Exploration Scientifique de la Mer Mediterranee, p. 241.

REYNOLDS, N. & WOOD, P. C. (1956). Improved techniques for the bacteriological examination of molluscan shellfish. *J. appl. Bact.*, **19**, 20.

SHERWOOD, H. P. & SCOTT THOMSON (1953). Bacteriological examination of shellfish as a basis for sanitary control. *Mon. Bull. Minist. Hlth*, **12**, 103.

SHEWAN, J. M. (1963). The differentiation of certain Gram negative bacteria frequently encountered in marine environments. In *Symposium on marine microbiology* (Oppenheimer, C. H., ed.). Springfield, Illinois: Charles C. Thomas.

WOOD, P. C. (1965). The effect of water temperature on the sanitary quality of *Ostrea edulis* and *Crassostrea angulata* held in polluted waters. *Pollutions marines par les micro-organismes et les produits petroliers*, Symposium de Monaco (avril 1964). Monaco: Commission Internationale pour l'Exploration Scientifique de la Mer Mediterranee, p. 307.

WOOD, P. C. (1969). *The production of clean shellfish.* Lab. Leafl. Fish. Lab. Burnham-on-Crouch, (N.S.) No. 20.

Techniques for the Study of Mixed Fouling Populations

T. LOVEGROVE

International Marine Coatings, Biological Laboratory, Newton Ferrers, Plymouth, England

Introduction

"Fouling results from the growth of animals and plants on the surface of submerged objects." This simple statement gives little indication of the scope of a problem which has troubled mankind for as long as the oceans have been used, with written records as early as the fifth century B.C. The organisms concerned include representatives of most of the invertebrate phyla as well as algae, fungi and bacteria; indeed one estimate suggests that about 2000 species are available (US Naval Institute, 1952).

The detrimental effects of fouling are wide, including decreasing water flow through pipes, the plugging of orifices, the addition of excessive weight, directly interfering with moving components and the most well known of all, that of increasing the resistance of ships' hulls. Houghton (1970) quotes an estimated £50 million per annum being spent in the United Kingdom in extra docking fees and other associated costs for ships, and in Norway the figure was about £10 000 a year for a 10 000-tonne ship. Tankers of 250 000 tonnes are in service and there are only four docks in the world capable of accommodating ships of this size and the cost of a week in dry dock is £100 000 to £150 000. It is therefore desirable that ships be docked as infrequently as possible especially since drydocking costs in Northern Europe have increased by up to 100% in the last five years. Between 1972 and 1975 fuel costs increased by nearly 300% and at present the fuel bill for a 250 000-tonne VLCC (Very Large Crude Carrier) would be about £1·6 million per annum. Clearly, even a small saving in fuel cost here can mean a great deal when every 1% saving is worth about £16 000. Using the current generation of high performance anti-fouling, such a vessel will have lost 0·9 knots or more in speed after 22 months in service with an average loss of about half a knot over the normal two year period between drydockings and this represents 2·5% loss in speed equal to £125 000 on fuel alone, without considering savings on drydockings and the loss of days trading.

Running costs of this magnitude have created an increasing demand for more effective antifouling measures both as regards biocidal activity and length of life. Many ideas have been tried including using chlorination devices, ultrasonics and radioactive materials, but the best method remains that of applying a paint containing a bioactive substance which slowly leaches out into the laminar layer of water in its immediate vicinity. The rate at which the pace of development has increased can be appreciated by a brief consideration of the history of anti-foulings. The first patent for a paint containing copper was granted in England in 1625 and up to the start of the last war in 1939 the need was for anti-fouling protection for only six months. The current generation of high-performance coatings will provide protection for about two years using various biocides. With the increasing technology of ship-borne equipment it has been established that there is a need for anti-fouling coatings capable of lasting for four years or more out of dock. Considerable research and development effort was required to produce the present situation and future improvement must be with the development of new paint matrices and the selection of new anti-foulants.

Anti-fouling Testing

A fouling community is based upon complex interactions between the various organisms present and their relationship to the physical habitat. The dominant organism, depending upon geography, time of year, substrate and biotic factors, may subsequently influence the future growth and attachment of other species. Because of the complexity of this actively changing environment it is only possible to carry out meaningful trials in a marine situation with a rich source of animal and plant fouling organisms.

Ideally, all tests should be carried out on ships' hulls but this is obviously impractical and therefore raft testing has been the accepted method whereby developmental work is carried out. The classical procedure involves suspending test panels vertically in the water from suitable attachments on the raft and whilst tests of this type can provide adequate information for general comparisons of anti-fouling performance, these methods are inadequate for more critical assessments. The main disadvantage in vertical plate testing is that because weed fouling can only grow in the presence of light, it is essential to space the panels far enough apart to receive adequate daylight, whilst barnacles and most animal fouling grow more readily when shaded from sunlight. To effect a compromise large or deep panels must be used, reaching from near the waterline to a depth which will serve to reduce the light. Such panels are unwieldy and have the second best qualities of such a compromise. In addition many biocides tend to be specific in action; for example, D.D.T. is only effective against barnacle fouling and thus paints based on D.D.T. immersed on vertical raft plates become so fouled with weed that barnacles have no room to settle and thus any test of barnacle resistance by the compound is nullified. It is however possible to take advantage of some of the factors influencing settlement and develop a technique which broadly separates the rich miscellany of fouling organisms.

On the hull of a ship the incidence and variety of fouling can best be considered by separating the surface liable to fouling into three main areas.

(1) The well-lit waterline liable to heavy fouling by weed.
(2) The sparsely illuminated flats where settlement of barnacles and animal fouling will predominate.
(3) The area between, centred on the turn-of-bilge, where light is still restricted but less so, and where both plant and animal fouling may occur.

To take advantage of such realistic conditions a new style testplate carrier was designed at the Newton Ferrers Laboratory which fulfils the necessary requirements while at the same time allowing easy access at any time for examination of any facet of exposure. (Sparrow, 1966).

The Turtle Plate Carrier

Essentially this carrier is in the form of a hollow wooden octagonal cylinder 5·5 m long and 1·2 m in diameter with the ends of the cylinder tapered to a point (Fig. 1). A central spine or spindle runs from end to end, protruding at each end to engage in a submerged bearing, so that the whole cylinder can be held underwater.

It is further held in position by removable chains between two parallel moored floats or pontoons 8 m in length which serve also as inspection platforms and are connected rigidly by strong 3·7 m long bridging pieces carrying the bearings at each end. The degree of submergence can be varied by a system for raising and lowering the bearings, but normally the highest point of the octagonal cylinder is immersed just below water level. The two uppermost facets are not normally used but may be utilized as storage areas when another raft is being serviced. The remaining six facets, three on each side, are immersed to simulate the waterlines, turn-of-bilge and flats areas of a ship (Fig. 2).

The topmost or waterline facet is tilted upwards so that it receives the

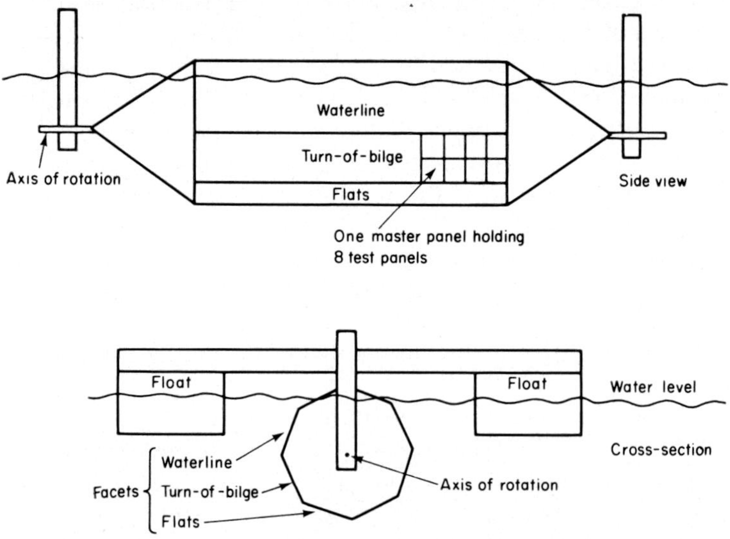

FIG. 1. Schematic layout of a Turtle raft.

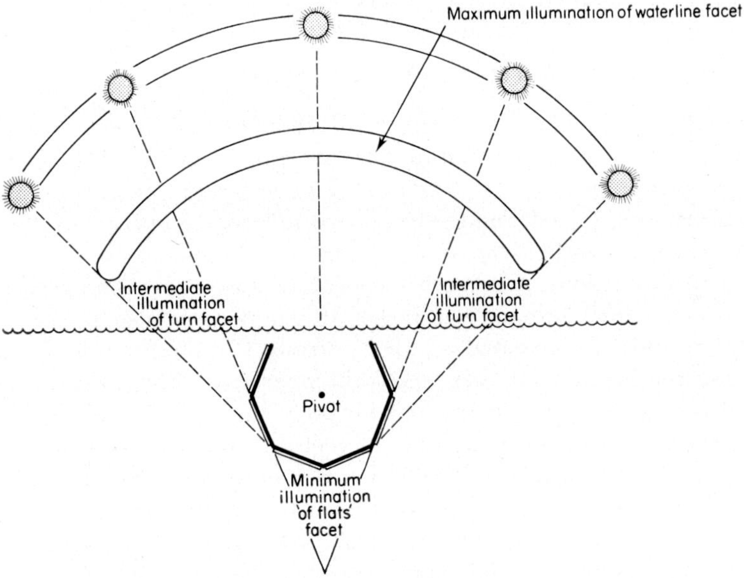

FIG. 2. Principle of Turtle action.

maximum light of all the facets and thus weed growth is stimulated. The lowermost flats facet receives the least light which provides conditions attractive to animal fouling but reduces algal growth to a minimum. The turn-of-bilge receives light intermediate between the extremes and resembles the conditions on a traditional vertically immersed plate so that both weed and animal settlements occur. The characteristic fouling of each of the three main areas of a ship's hull are thus reproduced in miniature and weed and animal fouling can be studied both independently and together (Fig. 3).

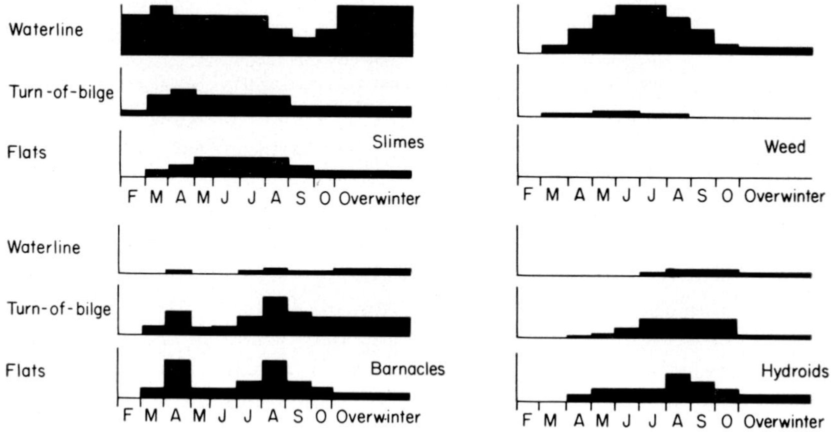

FIG. 3. Monthly settlement and distribution of the main fouling groups on a Turtle raft at Newton Ferrers, Devon.

Test panels which can be of any desired material are secured by fastening in groups to 12 master panels which are bolted to each of the submerged facets. The centre of the waterline facet master panel lies about 0·5 m below the surface, the turn-of-bilge panel about 1 m and the flats panel at about $1\frac{1}{2}$ m.

Inspection of panels can be carried out rapidly and simply by releasing the retaining chains, bringing the upper facets above the waterline and then rotating the cylinder to bring each facet into view for examination. The carrier is thus made to "turn turtle" for inspection and this led to its being known as a "Turtle raft". Experimental materials may be subjected to tests in triplicate, being applied on three panels which are immersed respectively in the three positions corresponding to waterline, turn-of-bilge and flats whilst controls similarly disposed are provided in the form of non-toxic areas and materials of known performance.

Uses of a Turtle Raft

Biological

Ideal conditions are provided for the study of settlement of both animal and plant fouling organisms with minimal interference one with the other. The development of dominance and the various seasonal factors can be readily demonstrated.

Primary screening for new bioactive substances

Over the past five years some 3000 compounds have been screened for their potential as anti-foulants and the Turtle method of testing has been particularly useful in that it shows up specific bioactivity against weed and animal fouling (Plate 1).

A small known quantity of the test compound is placed into a circular plastic container whose open end is then covered with a membrane of known porosity; the complete unit is attached to a Turtle master panel so that the material can diffuse freely into the free water layer around it. Regular inspections, recording the time, nature and quantity of settlement, allow a history of each such trial to be constructed. Each master panel holds 35 trial compounds together with 4 standard biocides and a non-toxic membrane for control purposes and duplicate panels are immersed on waterline and flats facets. Where only very small amounts are available then a single immersion on the turn-of-bilge facet is used.

At the end of the standard immersion period of 13 weeks the activity of the 35 trial materials can be compared with the 5 controls and compounds showing selective effects would be investigated further. Compounds showing some marginal control of fouling may suggest areas for further work such as analogue synthesis to improve performance. After completion of the membrane test, the contents of all containers are checked visually for material loss and an empty or nearly empty container indicates an unacceptably high solubility. Materials which have shown promise in this screening operation are then tested in the laboratory for their toxicity against algal spores and barnacle larvae, depending on whether their bioactivity is general or specific.

PLATE 1. Selective fouling action of a Turtle raft and its application to the screening of potential anti-foulants. (*left*) Waterline facet, (*right*) flats facet.

Secondary screening of new potential anti-foulants in paint media

Selected compounds are incorporated into paints to determine their performance in relation to the particular paint vehicle, the concentration of active ingredient, existing products for cost/performance evaluation. Only small amounts of experimental material are generally available and therefore small (100 cm^2) plaques are used.

Assessments of settled fouling are carried out which allow the compound's performance against the various forms of fouling at a particular concentration in a particular medium to be determined. A total of 32 such plaques per master panel are immersed in duplicate on waterline and flats facets. Where less refined trials are contemplated the turn-of-bilge facet is used to provide a general view of activity.

Developmental work

Large plates (400 cm^2) are used for development trials and eight plates fill a master panel. The paints in this case would be either variants on standard products or new products in an advanced stage of commercial development.

Each of the three facets on one side of the Turtle plate carrier can accommodate trials of a single type or mixtures thereof. The maximum capacity for trials of a particular type per facet of the whole carrier would be

(1) primary screening—480 trials;
(2) secondary screening—384 trials;
(3) developmental work—96 trials.

Two Turtle rafts have been in regular use at Newton Ferrers for the past 16 years and have been a great advance in marine testing techniques. They have been particularly useful in latter years in screening materials for their activity as anti-foulants since the Turtle method of testing shows up specific bioactivity against weed and animal fouling.

References

HOUGHTON, D. R. (1970). Marine Anti-fouling. *Underwat. Sci. Technol. J.*, **2**, 100.

SPARROW, B. W. (1966). Essais, sur radeux, de peintures de carenes de navires. *Peintures, Pigments, Vernis*, **42**, 4.

US NAVAL INSTITUTE (1952). *Marine fouling and its prevention.* Woods Hole Oceanographic Institute report to US Navy, Annapolis, Md.

A Technique for the Enumeration of Heterotrophic Nitrate-reducing Bacteria

R. W. Horsley

Freshwater Biological Association, The Ferry House, Ambleside, Cumbria, England

Introduction

In this paper details of a technique for the enumeration of heterotrophic nitrate-reducing bacteria, relative to the total heterotrophic bacterial population, within freshwater are presented.

Heterotrophic bacteria that are mesophilic aerobes, or facultative anaerobes, are often found in aquatic habitats which have widely varying temperatures and oxygen concentrations, and contain complex biodegradable organic matter. Their physiology and diverse nutritional requirements permit them to be isolated from a water sample on a variety of bacteriological media when incubated under either aerobic or anaerobic conditions. Nitrate-reducing heterotrophs, notably denitrifying strains, have been isolated and grown on complex bacteriological media formulated with combinations of peptone, beef and/or yeast extract, glucose, glycerol, sodium acetate, sodium succinate and mineral salts (Sacks and Barker, 1949, 1952; McNall and Atkinson, 1956; Society of American Bacteriologists, 1957; Stanier *et al.*, 1966; Abd-el-Malek *et al.*, 1974) and also on defined media in which either glucose (Marshall *et al.*, 1953) or, as in Giltay's medium (Fred and Waksman, 1928), sodium citrate and asparagine are the sole carbon sources. Ecological studies invariably require an understanding of the numerical relationship of one specific group of bacteria to that of another within a habitat. The successful enumeration of a specific group of organisms present within a habitat depends upon the exploitation of a biochemical reaction characteristic of the organisms sought. However, no single method exploiting a biochemical feature may be relied upon to enumerate all of the bacteria characterized by a certain biochemical reaction. A suitable technique for the enumeration of heterotrophs is unlikely to be suitable for enumerating autotrophs, primarily because of the different nutritional requirements of autotrophs.

The numbers of viable heterotrophic bacteria within a water sample may be determined by either the spread plate (Collins *et al.*, 1973) or roll-tube methods (Baker *et al.*, 1955), the membrane filtration (Mulvany, 1969) or Most Probable Number (MPN) techniques (Public Health and Medical Subjects Report, 1969). The first three techniques give no indication of the biochemical activities of the colonies counted, whilst the fourth may be used to determine the numbers of bacteria within a population that can effect a specific biochemical reaction. The MPN technique is often used for the enumeration of faecal coliforms and *Escherichia coli*, when the production of acid and gas from glucose in a nutrient broth is taken to indicate the presence of these bacteria.

Nitrate Reduction

Within a sample of freshwater, containing a mixed heterotrophic bacterial population growing aerobically, there will probably be three groups of organisms displaying a distinct reaction upon nitrate; namely, one group reducing nitrate to nitrite only, another reducing nitrate to beyond nitrite yielding ammonia as the end product and another failing to degrade the nitrate ion. A similar pattern may occur in a population growing anaerobically, except that the end product produced by organisms reducing nitrite will either be atmospheric nitrogen or nitrous oxide. The anaerobic reduction of nitrate to either of these end products is termed denitrification.

Members of the taxa Enterobacteriaceae, *Pseudomonas*, *Alcaligenes*, *Moraxella*, *Acinetobacter*, *Aeromonas*, *Vibrio*, *Flavobacterium*, *Cytophaga*, *Bacillus*, *Corynebacterium*, *Arthrobacter* and Micrococcaceae often represent the heterotrophic bacterial flora within freshwater (Collins, 1963; Horsley, 1973). According to Bergey (1974) many species belonging to these taxa can reduce nitrate, certain strains classified as *Pseudomonas Cytophaga* and *Bacillus* notably being capable of denitrification; the members of the other taxa listed above often reduce nitrate to nitrite only, or under aerobic conditions, to ammonia.

The Enumeration of Heterotrophic Nitrate-reducing Bacteria

A technique that would permit the rapid enumeration of the two categories of organisms displaying a reducing action upon nitrate, and yet be practical for use in an ecological survey of the bacterial denitrification processes within freshwater, was sought.

Initially, the development of a technique similar to that employed in the oxidase test of Kovacs (1956) was attempted. It was hoped that colonies transferred from an agar plate and applied directly to filter paper soaked in a solution of either nitrate or nitrite would reduce these ions. Unfortunately, definite results could not be obtained by this method, nor by testing colonies direct on an agar plate for the reduction of these ions.

A most probable number technique was considered to be unsuitable for the enumeration of heterotrophic nitrate-reducing bacteria, because it would not allow the differential enumeration of organisms characterized by the three possible reactions upon nitrate. The most probable number technique depends upon adding measured volumes of the water to be tested, or dilutions of it, to tubes containing a liquid medium. It is assumed that one or more organisms inoculated into the medium will show growth and a biochemical reaction that is characteristic of the organisms sought. Provided that negative results can be recorded for some tubes, the most probable number of organisms in the original sample can be estimated from the number of tubes giving a positive reaction and statistical tables of probability. The most probable number of nitrate-reducing bacteria determined to be present within a water sample, would possibly be those representing a nitrate-reducing group numerically dominant at a specific time; the technique would not reveal those members of a group present in lesser numbers.

Recently, Patriquin and Knowles (1974) have shown that higher numbers of denitrifying bacteria were estimated by selecting colonies from an agar plate than by techniques dependent upon the production of gas and the disappearance of nitrite as the criterion for positive counts. However, when the production of N_2O was the criterion for a positive count the results for the two methods were similar.

The proposed method

The following technique, although labour intensive, was developed for the enumeration of nitrate-reducing heterotrophs and was considered to be suitable for use in an ecological survey of the bacteria participating in the denitrification processes within Grasmere.

Grasmere is an eutrophic freshwater lake in the English Lake District; whose typical seasonal patterns, at 20 m depth, for temperature, percentage oxygen saturation and $NO_3 + NO_2$—N are as presented in Fig. 1.

The Casein-Peptone-Starch (CPS) agar medium developed by Collins (Collins and Willoughby, 1962) was selected for the determination of the total viable heterotrophic count, and the nitrate medium of Stanier *et al.*

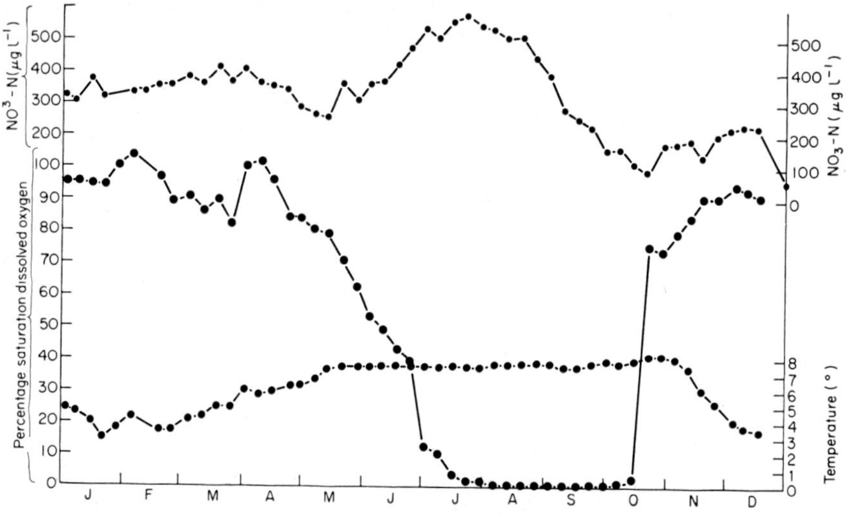

FIG. 1. Seasonal determinations for dissolved oxygen, temperature and NO_3 nitrogen in Grasmere at 20 m depth.

(1966) to determine the nitrate-reducing ability of colonies taken from the total viable count plate. Both of these media were modified by the addition of KNO_3 at a concentration of either 0·01% or 0·1% (w/v).

The CPS agar was chosen for the determination of the total numbers of heterotrophs because besides the studies of Collins and Willoughby (1962), those of Staples and Fry (1973) have indicated that it is a suitable medium for the enumeration of heterotrophic bacteria from freshwater.

The medium of Stanier *et al.* (1966) was selected because it is a semi-defined medium suitable for controlled modification, particularly of the carbohydrate substrate. It was also believed that most strains of heterotrophic bacteria would grow in this medium. Furthermore, even though formulated with a higher concentration of KNO_3, this medium had been used by Stanier *et al.* (1966) in studies concerning nitrate reduction by strains of *Pseudomonas*, organisms commonly found in freshwater. Abd-el-Malek *et al.* (1974) observed that a medium containing peptone and yeast extract gave maximum MPN counts of denitrifiers from soils, whilst a medium of composition similar to that of Stanier *et al.* gave counts ten times lower.

Giltay's medium (Fred and Waksman, 1928), with sodium citrate and asparagine as the carbon sources, was considered to be unsuitable for determining the nitrate-reducing ability of the isolates, because certain *Flavobacterium* and *Cytophaga* strains which often compose a part of the bacterial flora of water are reported to be able to reduce nitrate but unable to utilize citrate as a sole carbon source (Bergey, 1974).

Studies employing pure cultures have indicated that the modified nitrate broth of Stanier *et al.* was slightly superior to a casein-peptone-starch broth (Collins and Willoughby, 1962) prepared with KNO_3, for determining the nitrate-reducing ability of the selected colonies.

Further, it was observed that a liquid medium facilitated the assessment of the nitrate reduction reaction of the organisms isolated. The reagents employed to confirm the presence of nitrite produced a coloured complex more rapidly with a broth than an agar medium, particularly when cadmium metal was used to determine the presence of nitrate. The agar possibly restricted the diffusion of the reagents.

The CPS agar plates were prepared in the standard manner (Baker *et al.*, 1955). The nitrate broth was dispensed in 2·0 ml volumes into each of the squares within a Replidish (Sterilin Ltd, 43–45 Broad St, Teddington, Middlesex TW11 8QZ, England). Pre-inoculation contamination of these media, particularly of the Replidishes, by moulds was eliminated when they were dispensed in a vertical flow clean air cabinet (Pathfinder, Environmental, 1974 Ltd, Solent Rd, Havant, Hampshire PO9 1JF, England).

Total viable count

A 0·1 ml volume of water sample, or serially diluted sample, was spread aseptically over the surface of a CPS agar plate in a standard manner (Collins *et al.*, 1973). To obtain a statistically reliable total viable count five agar plates were inoculated in the same manner. A count of between 30 and 100 colonies per agar plate was found to be suitable for enumeration, also for the selection and transfer of pure cultures into the nitrate broth. When water samples from Grasmere were diluted 1:10 in sterile Ferry House well water sufficient separated colonies grew on the agar plates when they were incubated aerobically. Grasmere water samples were not diluted when the incubation atmosphere was to be anaerobic.

To obtain a differential count of heterotrophs able to reduce nitrate under aerobic and anaerobic atmospheres, two sets of CPS agar plates were prepared. One set of plates prepared with 0·1% KNO_3 (w/v) in the medium, was incubated aerobically. After cutting two slots in the polystyrene Petri dish with either an electrically heated wire or a heated spatula, the other set, containing 0·01% KNO_3 (w/v) in the medium, was incubated in an anaerobic jar (Baird and Tatlock Ltd, Chadwell Heath, Essex, England) filled, according to the regime of Collee *et al.* (1972), with an H_2/CO_2 mixture. This gas mixture was either 95% H_2/5% CO_2 (v/v) as specially supplied by British Oxygen Co. Ltd, Special Gases (Deer Park Rd, London SW19 3UF, England) or that derived from a GasPak (Becton, Dickinson, UK, Ltd, York House, Empire Way, Wembley, London HA9 0PS, England). The slots in the Petri dish ensured the removal of air from within the dish, and its substitution with the H_2/CO_2 mixture.

The decision to reduce the KNO_3 concentration in the media to be incubated anaerobically, was caused by the results of comparative experiments employing 0·01% and 0·1% KNO_3 (w/v) in both the agar and broth media. The details and results of these experiments are presented later in this paper.

The nitrate-reducing ability of the heterotrophic bacterial population present in the water sample was determined by randomly selecting 25 colonies from a count plate, and inoculating them into the nitrate broth medium. A CPS agar plate with colonies derived from the water sample was divided into 18 numbered squares, and by using random number tables to select the squares in sequence, 25 colonies were picked from within them. The preparation of a randomly numbered template obviated delineating squares on the back of an agar plate, also constant reference to the random number tables. This technique was adopted in an endeavour to avoid biased selection of the colonies. The colonies were picked from the agar plates with sterile, wooden cocktail sticks and inoculated directly into the nitrate broth contained within a Replidish square. Cultures previously incubated aerobically were inoculated into broth prepared with 0·1% KNO_3 (w/v); broth containing 0·01% KNO_3 (w/v) was used for those cultures that had been incubated anaerobically. This inoculating procedure was carried out in a vertical flow clean air cabinet, which allowed the Replidish to remain open whilst the squares were inoculated.

The inoculated Replidish was incubated at 20° for 7 days in an atmosphere corresponding to that under which the cultures had been grown. When incubated anaerobically four Replidishes could be accommodated within an anaerobic jar lying horizontally in a simple wooden cradle. The horizontal use of an anaerobic jar has been described by Jayne-Williams (1975).

An incubation temperature of 20° was selected because it was believed that the majority of heterotrophic bacteria within Grasmere would be mesophiles and facultative psychrophiles. Also, as the seasonal temperature range within the water column is often from 3 to 20° it was considered that 20° would be favourable for their growth.

Detection of the nitrite ion

After the incubation period the Replidish squares were carefully examined to ascertain whether or not the inoculum had grown, following which the reagents to determine the presence of nitrite and, when appropriate, nitrate were added.

There were invariably a few isolates that failed to grow in the nitrate broth after transfer from the CPS agar plate. The failure of organisms to grow after initial isolation has been reported by Kriss (1963) for isolates from the Greenland Sea.

The Griess-Ilosvay reagent (Ilosvay, 1889; Wilson and Miles, 1946) is no longer favoured for determining the presence of nitrite; α-naphthylamine, the reagent that couples with diazotized sulphanilic acid, is regarded as carcinogenic (Chester Beatty Research Institute, 1966). The reagent of Follet and Ratcliff (1963) is presently preferred for use in bacteriological nitrate reduction analysis (Spencer, 1969); it produces an orange coloured complex with nitrite and is suitable for use in the presence of nitrate concentrations greater than 10 mg litre^{-1}. Thus this reagent can be used in the presence of 0·1% (w/v) KNO_3 (NO_3—N 138·46 mg litre^{-1}). The more sensitive reagent of Elliott and Porter (1971), which produces a magenta coloured complex with nitrite, is more suitable in the presence of 0·01% (w/v) KNO_3. Two or three drops of the reagents are added to the Replidish cultures, using either a Pasteur pipette or a Zipette (Jencons, Scientific Ltd, Mark Road, Hemel Hempstead, Hertfordshire, England).

After the addition of the appropriate nitrite reagent to the cultures, the number of cultures producing a coloured complex, indicating the reduction of nitrate to nitrite only, were recorded. Subsequently, a spatula tip of cadmium metal was added to those that did not produce a coloured complex. After allowing about 20 min to elapse, the number of cultures yielding a colour was recorded as not having reduced the nitrate. The cadmium metal reduced the undegraded nitrate to nitrite only. The remainder of the cultures not reacting with the nitrite reagent were recorded as having reduced the nitrate to possibly either NH_3 or N_2, depending upon the incubation atmosphere.

These data were related to the total viable count, and used to calculate the numbers of heterotrophic bacteria, expressed logarithmically, within the water sample, that could reduce nitrate to nitrite only, reduce it beyond nitrite or failed to degrade this ion.

Results of Studies Employing the Enumeration Technique

Selection of the KNO_3 concentration

The decision to use 0·01% and 0·1% KNO_3 (w/v) in the media was based upon the results of the studies of Bollag *et al.* (1970), also the observation that 0·1% KNO_3 has often been used in nitrate media (Society of American Bacteriologists, 1957; Abd-el-Malek *et al.*, 1974), and the results of a study comparing the effect of these KNO_3 concentrations in the enumeration media.

To ascertain the most suitable KNO_3 concentration to employ in the media, the numbers of heterotrophic nitrate-reducing bacteria within Grasmere water samples were determined in media prepared with 0·01% also 0·1% KNO_3 (w/v) that were incubated both aerobically and anaerobically. The criteria applied to determine the most suitable KNO_3 concentration were that the media should allow the majority of organisms to grow and reduce nitrate to beyond nitrite; failing this they should demonstrate maximal numbers reducing nitrate to nitrite only. The results of this study (Tables 1 and 2) indicated that a concentration of 0·1% KNO_3 (w/v) in the media was suitable for aerobic enumeration; under an anaerobic regime, 0·01% KNO_3 (w/v) was seemingly the superior concentration.

The random selection of colonies

The selection of 25 colonies only from the count plates was conditioned by the practical necessity to select the minimum number of colonies to represent the heterotrophic bacterial population in a water sample. The validity of this selection was confirmed by statistical analysis of the results obtained from Grasmere water samples.

Subsamples of 25 colonies were randomly selected, using a grid of random numbers, from an agar plate upon which the total viable count of heterotrophs in a water sample had been determined. It was necessary to ensure that a subsample of 25 colonies provided results which were both reproducible, and representative of the whole population. Both of these assumptions were checked by employing the χ^2 test. Six sets of 25 colonies derived from water at 6 m depth, and six sets from a 20 m sample were tested for nitrate reduction. No significant difference between the subsamples was found, and therefore the results were reproducible. The assumption that the selection of 25 colonies provides a result which is representative of the whole population, was checked by comparing the proportions of positive and negative isolates with those obtained when all the colonies on the plate were tested. Analysis of the results by the χ^2 test showed that such a selection was representative ($p < 0{\cdot}01$) in 75% of the samples analysed. Given the errors associated with spread-plate techniques, and the greatly increased labour involved in testing a higher proportion of the colonies, it was concluded that the enumeration procedure provided a reasonable estimate of the nitrate-reducing organisms in the heterotrophic population.

Results of a survey

A survey of the heterotrophic bacteria involved in the denitrification processes within Grasmere at 20 m depth, was undertaken in 1974. It was believed that this survey would confirm the practical use of the technique, and possibly its suitability for enumerating heterotrophic nitrate-reducing bacteria from freshwater. The results of this survey are presented in Figs 2 and 3. Details of the typical dissolved oxygen concentration, temperature and $NO_3 + NO_2$ –N concentration within Grasmere at 20 m depth are presented in Fig. 1.

The total numbers of viable heterotrophic bacteria as determined aerobically were often 10 times greater than those determined under an H_2/CO_2 atmosphere; there were 10^3–10^4 organisms ml^{-1} aerobically and 10–10^3 ml^{-1} anaerobically, with greater numbers tending to prevail in the periods January–April and October–December.

The enumeration technique indicated that within the heterotrophic bacterial population present at 20 m depth during April to October, there were marked variations in the numbers of nitrate-reducing organisms. The total heterotrophic bacterial population enumerated aerobically was almost numerically constant during this period. However, within this population there were seemingly three distinct cycles of increase and decline in the numbers of organisms representing the two nitrate-reducing groups. These cycles appear to be associated with the nitrate and dissolved oxygen concentrations at 20 m depth. Heterotrophs reducing nitrate to nitrite only, and nitrate to possibly ammonia were present in similar numbers during April and May; those reducing nitrate to nitrite only predominated throughout the second cycle that appeared within June and July. Similarly, during the months of August to October, inclusive, heterotrophs reducing nitrate to nitrite only were more numerous than those reducing nitrate to possibly ammonia and displaying a cycle during September only.

When the heterotrophic nitrate-reducing bacterial population was enumerated anaerobically, the variations in the numbers of nitrate-

TABLE 1. The effect of KNO_3 addition upon the quantitative analysis of nitrate reduction under aerobic conditions

		0·01% KNO_3 in the media			
Sample	TVC ml^{-1}*	$NO_3 \rightarrow NO_2$	Nitrate reduction $NO_3 \rightarrow NH_3$	No reduction	No growth
29.4.74					
0 m depth	4·259	3·463	3·162	3·764	3·903
6 m depth	2·973	2·478	2·177	2·052	2·575
12 m depth	2·380	1·602	1·301	2·079	1·903
20 m depth	3·568	3·124	2·170	3·073	2·948
6.5.74					
0 m depth	4·100	3·932	Nil	2·401	3·179
6 m depth	4·053	3·801	2·956	3·257	3·354
12 m depth	3·536	2·837	2·439	3·342	2·439
20 m depth	3·768	2·972	2·671	3·324	3·370
13.5.74					
0 m depth	4·445	3·348	Nil	3·650	4·326
6 m depth	2·732	1·811	Nil	2·447	2·288
12 m depth	2·913	2·214	Nil	2·214	2·692
20 m depth	4·082	2·683	Nil	3·529	3·872
20.5.74					
0 m depth	4·442	Nil	3·044	3·521	4·367
6 m depth	2·415	1·415	Nil	3·017	3·113
12 m depth	3·158	2·663	Nil	2·538	2·801
20 m depth	2·579	0·914	0·136	1·090	0·708
27.5.74					
0 m depth	2·662	2·410	Nil	2·219	Nil
6 m depth	3·225	2·304	Nil	4·003	2·672
12 m depth	2·778	2·194	1·416	2·530	1·893
20 m depth	3·120	2·324	2·199	2·625	2·722
3.6.74					
0 m depth	3·152	2·453	2·231	2·532	2·795
6 m depth	3·809	3·110	3·365	3·314	2·888
12 m depth	2·875	2·477	2·778	2·556	2·477
20 m depth	2·962	2·865	3·163	2·644	2·410

* TVC, total viable count; the mean of the count from five agar plates, expressed as Log_{10} of count 1·0 ml^{-1} of lake water.

	0·1% KNO_3 in the media					
	Nitrate reduction		No	No	Most favourable	
TVC ml^{-1}*	$NO_3 \rightarrow NO_2$	$NO_3 \rightarrow NH_3$	reduction	growth	KNO_3 concentration	
4·240	2·842	3·319	3·686	3·988	Either	
2·880	2·261	2·084	2·181	2·482	Either	Either KNO_3
2·342	2·000	1·301	1·602	1·778	Either	concentration
3·418	2·622	2·321	2·974	3·020	Either	
4·056	4·001	3·136	Nil	Nil	0·1%	
4·016	3·873	3·091	3·220	Nil	Either	Either KNO_3
3·528	2·431	2·732	3·276	2·829	Either	concentration
3·718	3·320	3·165	3·223	Nil	0·1%	
4·131	3·432	3·033	3·209	3·909	0·1%	
2·579	1·903	1·903	1·903	2·146	0·1%	0·1% KNO_3
3·278	2·675	Nil	2·199	3·102	Either	concentration
4·002	2·604	3·103	3·103	3·859	0·1%	
4·212	Nil	2·814	3·659	4·037	Either	
3·000	Nil	1·602	2·301	2·880	0·1%	0·1% KNO_3
3·255	2·158	2·158	3·003	2·702	0·1%	concentration
2·662	1·602	2·204	2·301	1·778	0·1%	
2·819	2·590	2·199	2·375	1·722	0·1%	
2·806	1·709	1·885	2·487	2·311	0·1%	Either KNO_3
2·792	1·871	1·871	2·508	2·172	Either	concentration
3·084	1·686	2·465	2·532	2·728	Either	
3·346	Nil	2·249	2·851	3·124	0·01%	
3·365	1·967	Nil	2·745	3·222	0·01%	0·01% KNO_3
2·793	1·872	Nil	2·509	2·350	0·01%	concentration
2·965	Nil	Nil	2·868	2·266	0·01%	

TABLE 2. The effect of KNO_3 addition upon the quantitative analysis of nitrate reductions under anaerobic conditions

0·01% KNO_3 in the media					
		Nitrate reduction		No	No
Sample	TVC ml^{-1}*	$NO_3 \rightarrow NO_2$	$NO_3 \rightarrow N_2$	reduction	growth
29.4.74					
0 m depth	0·301	$\bar{1}$·778	0·079	Nil	$\bar{1}$·301
6 m depth	1·857	1·237	1·635	Nil	1·061
12 m depth	0·301	Nil	0·301	Nil	Nil
20 m depth	0·602	Nil	0·602	Nil	Nil
6.5.74					
0 m depth	1·380	0·778	1·255	Nil	Nil
6 m depth	1·699	0·903	1·623	Nil	Nil
12 m depth	0·903	Nil	0·505	Nil	0·681
20 m depth	1·079	0·602	0·903	Nil	Nil
13.5.74					
0 m depth	1·763	0·666	1·727	Nil	Nil
6 m depth	1·699	0·602	1·662	Nil	Nil
12 m depth	0·301	0·301	Nil	Nil	Nil
20 m depth	1·716	0·920	1·640	Nil	Nil
20.5.74					
0 m depth	1·643	0·301	1·623	Nil	Nil
6 m depth	2·599	1·202	2·582	Nil	Nil
12 m depth	0·778	0·602	0·301	Nil	Nil
20 m depth	1·415	Nil	1·342	0·301	0·301
27.5.74					
0 m depth	0·778	0·722	$\bar{1}$·142	$\bar{1}$·619	Nil
6 m depth	2·008	1·889	1·388	Nil	Nil
12 m depth	1·531	Nil	1·531	Nil	Nil
20 m depth	1·380	0·602	1·301	Nil	Nil
3.6.74					
0 m depth	1·146	0·903	0·778	Nil	Nil
6 m depth	1·204	0·602	1·079	Nil	Nil
12 m depth	1·204	Nil	1·146	Nil	0·301
20 m depth	1·380	Nil	1·380	Nil	Nil

* TVC, total viable count; the mean of the count from five agar plates, expressed as Log_{10} of count 1·0 ml^{-1} of lake water.

	$0{\cdot}1\%$ KNO_3 in the media					
	Nitrate reduction					
TVC ml^{-1}*	$NO_3 \rightarrow NO_2$	$NO_3 \rightarrow N_2$	No reduction	No growth	Most favourable KNO_3 concentration	
1·079	1·000	$\bar{1}$·301	Nil	Nil	0·1%	
1·748	1·730	0·350	Nil	Nil	0·1%	Either KNO_3
1·000	1·000	Nil	Nil	Nil	0·01%	concentration
0·602	0·602	Nil	Nil	Nil	0·01%	
1·204	1·204	Nil	Nil	Nil	0·01%	
1·857	1·801	0·936	Nil	Nil	0·01%	0·01% KNO_3
1·000	1·000	Nil	Nil	Nil	0·01%	concentration
1·447	1·447	Nil	Nil	Nil	0·01%	
2·000	2·000	2·000	Nil	Nil	0·01%	
1·869	1·851	1·851	0·471	Nil	0·01%	0·01% KNO_3
1·415	1·342	1·342	0·602	Nil	0·1%	concentration
1·869	1·813	1·813	0·948	Nil	0·01%	
1·869	2·659	Nil	Nil	Nil	0·01%	
2·411	2·411	Nil	Nil	Nil	0·01%	0·01% KNO_3
1·301	1·301	Nil	Nil	Nil	0·01%	concentration
1·819	1·819	Nil	Nil	Nil	0·01%	
1·716	1·716	Nil	Nil	Nil	0·01%	
2·230	2·133	1·531	Nil	Nil	Either	0·01% KNO_3
2·204	2·204	Nil	Nil	Nil	0·01%	concentration
2·230	2·230	Nil	Nil	Nil	0·01%	
1·505	1·447	0·301	Nil	0·301	Either	
1·662	1·543	0·565	Nil	0·866	0·01%	0·01% KNO_3
1·342	1·204	0·301	Nil	0·602	0·01%	concentration
1·477	1·447	Nil	Nil	0·301	0·01%	

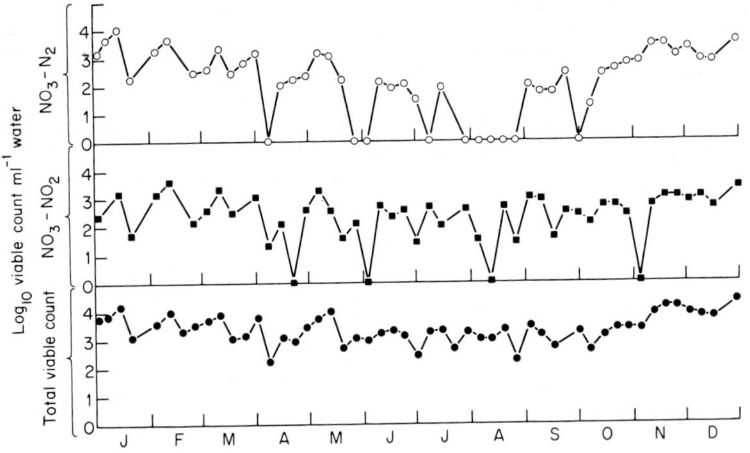

FIG. 2. Counts of heterotrophic bacteria reducing nitrate under aerobic conditions in Grasmere at 20 m depth during 1974.

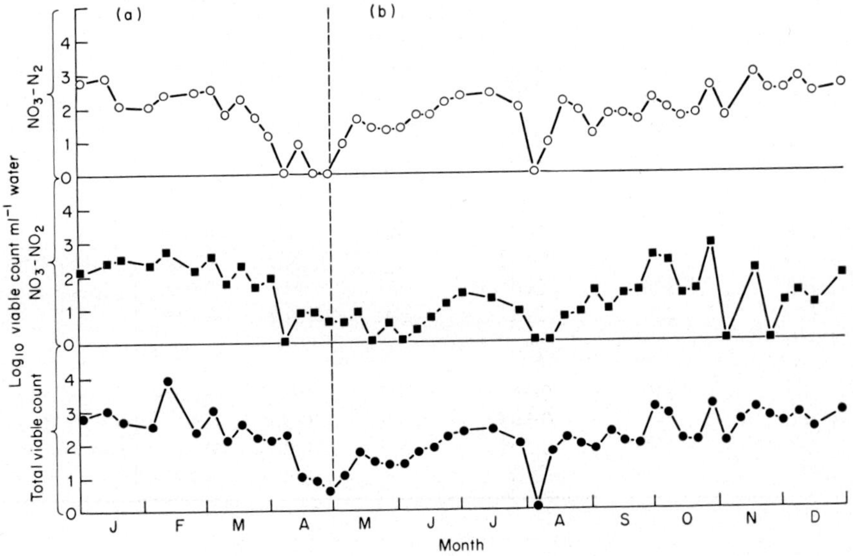

FIG. 3. Counts of heterotrophic bacteria reducing nitrate under anaerobic conditions in Grasmere at 20 m depth during 1974. (a) 2nd January to 29th April with 0·1% KNO_3 in medium. (b) With 0·01% KNO_3 in medium.

reducing organisms within the population were less marked than were observed aerobically. Throughout the year nitrate-reducing organisms composed almost the whole of the total heterotrophic bacterial population as determined anaerobically. During April this population declined from 10^3 to less than 10 organisms ml^{-1}, those strains reducing nitrate to nitrite only tending to be most numerous. Within the period May to July, inclusive, organisms reducing nitrate to possibly gaseous nitrogen displayed a pattern of increase and decline in numbers, which, because they formed the major component, was reflected by the total population. Throughout these three months heterotrophs reducing nitrate to nitrite only were represented in lesser numbers, but during June and July they displayed a similar numerical pattern.

The results of this survey have indicated that the technique was suitable, and practical for the enumeration of those heterotrophic nitrate-reducing bacteria that were present in Grasmere at 20 m depth. However, the technique does not indicate that the nitrate-reducing heterotrophs were actively reducing the nitrate present within the lake waters.

Conclusion

A technique has been developed that will permit the enumeration of the heterotrophic bacteria within a water sample that are able to reduce nitrate to nitrite only, also nitrate to beyond nitrite, relative to the total numbers of heterotrophs present.

A survey of the heterotrophic nitrate-reducing bacterial population within Grasmere has indicated that this technique is suitable for ecological surveys.

This enumeration technique, with the media suitably modified, will possibly be suitable for the enumeration of heterotrophic nitrate-reducing bacteria from marine and estuarine waters, pickling brines and soil, besides other environments in which nitrate is present.

Acknowledgements

I acknowledge the help and encouragement of Vera G. Collins throughout this study; also the advice on statistical analysis from Dr. J. M. Elliott, and the technical assistance of Mrs. José Roscoe.

This study was a part of the research programme of the Freshwater Biological Association.

Part of this study was funded by the Department of the Environment under Contract DGR 480/32.

References

Abd-el-Malek, Y., Hosny, I. & Eman, N. F. (1974). Evaluation of media, used for enumeration of denitrifying bacteria. *Zentbl. Bakt. ParasitKde* (*Abt II*), **129**, 415.

Baker, F. J., Silverton, R. E. & Luckcock, E. D. (1955). In *An introduction to medical laboratory technology* (Baker, F. J., Silverton, R. E. & Luckcock, E. D., eds). London: Butterworth.

Bergey (1974). *Bergey's manual of determinative bacteriology*. 8th edn (Buchanan, R. E. & Gibbons, N. E., eds). Baltimore: Williams and Wilkins.

Bollag, J. M., Orcutt, M. L. & Bollag, B. (1970). Denitrification by isolated soil bacteria under various environmental conditions. *Proc. Soil Sci. Soc. An.*, **34**, 875.

Chester Beatty Research Institute (1966). *Precautions for laboratory workers who handle carcinogenic aromatic amines*. London: Chester Beatty Research Institute.

Collee, J. G., Watt, B., Fowler, E. B. & Brown, R. (1972). An evaluation of the Gaspak system in the culture of anaerobic bacteria. *J. appl. Bact.*, **35**, 71.

Collins, V. G. (1963). The distribution and ecology of bacteria in freshwater. *Wat. Treat. Exam.*, **12**, 40.

Collins, V. G. & Willoughby, L. G. (1962). The distribution of bacteria and fungal spores in Blelham Tarn with particular reference to an experimental overturn. *Arch. Mikrobiol.*, **43**, 294.

Collins, V. G., Jones, J. G., Hendrie, M. S., Shewan, J. M., Wynne-Williams, D. D. & Rhodes, M. E. (1973). Sampling and estimation of bacterial populations in the aquatic environment. In *Sampling—microbiological monitoring of environments* (Board, R. G. & Lovelock, D. W., eds). Soc. appl. Bact. Tech. Ser. No. 7. London and New York: Academic Press, p. 77.

Elliott, R. J. & Porter, A. G. (1971). A rapid cadmium reduction method for the determination of nitrate in bacon and curing brines. *Analyst*, **96**, 522.

Follett, M. J. & Ratcliffe, P. W. (1963). Determination of nitrite and nitrate in meat products. *J. Sci. Fd. Agric.*, **14**, 138.

Fred, E. B. & Waksman, S. A. (1928). *Laboratory manual of general microbiology, with special reference to the microorganisms of the soil*. 1st edn. New York and London: McGraw-Hill.

Horsley, R. W. (1973). The bacterial flora of the Atlantic salmon (*Salmo salar* L.) in relation to its environment. *J. appl. Bact.*, **36**, 377.

Ilosvay de N. Ilosva, M. L. (1889). L'acide azoteux dans la salive et dans l'air exhalé (VI). *Bull. Soc. Chem. Paris*, **2**, 388.

Jayne-Williams, D. J. (1975). Miniaturized methods for the characterization of bacterial isolates. *J. appl. Bact.*, **38**, 305.

Kovacs, M. (1956). Identification of *Pseudomonas pyocyanea* by the oxidase reaction. *Nature, Lond.*, **178**, 703.

Kriss, A. E. (1963). Distribution of heterotrophic microorganisms throughout the seas and oceans (Species growing on laboratory media). In *Marine microbiology* (*deep sea*) (trans. by Shewan, J. M. & Kabata, Z.). Edinburgh and London: Oliver and Boyd.

McNall, E. G. & Atkinson, D. E. (1956). I. Growth of *Escherichia coli* with nitrate as sole source of nitrogen. *J. Bact.*, **72**, 226.

MARSHALL, R. O., DISHBURGER, H. J., MACVICAR, R. & HALLMARK, G. D. (1953). Studies on the effect of aeration on nitrate reduction by *Pseudomonas* species using N^{15}. *J. Bact.*, **66**, 254.

MULVANY, J. G. (1969). Membrane techniques in microbiology. In *Methods in microbiology*, Vol. I (Norris, J. R. & Ribbons, D. W., eds). London and New York: Academic Press, p. 205.

PATRIQUIN, D. G. & KNOWLES, R. (1974). Denitrifying bacteria in some shallow water marine sediments: enumeration and gas production. *Can. J. Microbiol.*, **20**, 1037.

PUBLIC HEALTH AND MEDICAL SUBJECTS REPORT NO. 71 (1969). *The bacteriological examination of water supplies*, 4th edn. London: Her Majesty's Stationery Office.

SACKS, L. E. & BARKER, H. A. (1949). The influence of oxygen on nitrate and nitrite reduction. *J. Bact.*, **58**, 11.

SACKS, L. E. & BARKER, H. A. (1952). Substrate oxidation and nitrous oxide utilization in denitrification. *J. Bact.*, **64**, 247.

SOCIETY OF AMERICAN BACTERIOLOGISTS (1957). In *Manual of microbiological methods*. Committee on bacteriological technic. New York and London: McGraw-Hill.

SPENCER, R. (1969). A new procedure for determining the ability of micro-organisms to reduce nitrate and nitrite. *Lab. Practice*, **18**, 1286.

STANIER, R. Y., PALLERONI, N. J. & DOUDOROFF, M. (1966). The aerobic Pseudomonads: a taxonomic study. *J. gen. Microbiol.*, **43**, 159.

STAPLES, D. G. & FRY, J. C. (1973). A medium for counting aquatic heterotrophic bacteria in polluted and unpolluted waters. *J. appl. Bact.*, **36**, 179.

WILSON, G. S. & MILES, A. A. (1946). In *Topley and Wilson's principles of bacteriology and immunity*, 3rd edn (revised by Wilson, G. S. & Miles, A. A.). London: Edward Arnold, p. 368.

Analysis of the Avian Intestinal Flora

Ella M. Barnes, G. C. Mead, C. S. Impey and B. W. Adams

ARC Food Research Institute, Colney Lane, Norwich, Norfolk, England

Introduction

Many different kinds of anaerobic and facultatively anaerobic bacteria are normally found in the intestine of the healthy bird. The significance of this flora to the host in terms of nutrition, resistance to infection and various stress conditions has been the subject of considerable speculation and research (Jayne-Williams and Fuller, 1971; Barnes, 1975) but cannot be resolved until all the major components of the flora have been isolated and identified.

From the microbiologist's standpoint, the intestine can be divided into three sections: the duodenum and small intestine, where the numbers of bacteria are relatively low, generally less than 10^8 g^{-1}; the caeca, where a considerable microbial fermentation occurs, the number of bacteria present being approximately 10^{11} g^{-1} (wet weight) and the large intestine, which in most birds is relatively short and includes organisms from both the small intestine and caeca.

Prior to hatching, the intestinal tract of the chick is usually sterile and the intestinal flora is derived exclusively from the environment. Within a few hours of hatching, faecal streptococci, enterobacteria and sometimes clostridia, can be found multiplying in the caeca and scattered randomly through the rest of the alimentary tract. High levels of faecal streptococci and enterobacteria persist for several days in the duodenum and small intestine; the lactobacilli become established only by about the third day, but by the seventh day they have almost completely replaced the other bacteria. Faecal streptococci are often found with the lactobacilli in the small intestine but very high numbers of *coli-aerogenes* bacteria or clostridia are unusual. The caecal flora takes several weeks to develop. Although 10^{10}–10^{11} bacteria g^{-1} are present from the first day onwards the initial flora was shown to consist almost entirely of faecal streptococci and *coli-aerogenes* bacteria with lactobacilli slowly increasing in numbers. After about the fourth day all three groups are gradually replaced by an anaerobic flora, their numbers falling to below 10^9 g^{-1} (Mead and Adams, 1975). The caecal flora which is predominantly composed of obligate anaerobes continues to change for several weeks and shows increasing complexity (Barnes *et al.*, 1972).

In analysing the microflora three approaches are used.

(1) The enumeration and isolation of the major components using both aerobic and anaerobic techniques, combined with selective and non-selective media.

(2) The isolation of organisms with properties of possible ecological significance, e.g. the uric acid-decomposing bacteria which may have a role in relation to caecal function.

(3) The detection of potentially pathogenic bacteria which may occur in very low numbers in the intestine, e.g. the salmonellae.

Organisms which are particularly difficult to isolate are obligate anaerobes which occur in the caeca at about 10^8 g^{-1} or lower and for which there are no selective media available at the present time. Amongst these are some characteristically very large organisms which can be seen by direct microscopical examination of the caecal contents. The finding of methane in some caeca (Shrimpton, 1966) indicates the presence of methane bacteria but these also have yet to be isolated and studied.

Media and Incubation Conditions

For use with the strictly anaerobic technique of Hungate (1950)

Anaerobic Dilution Solution (ADS) (Bryant and Burkey, 1953)
Supplemented Medium 10 (SM10) Agar

Details of the preparation of Medium 10 (Caldwell and Bryant, 1966) and its supplementation with liver and chicken faecal extract have been described by Barnes and Impey (1974). The roll-tubes are incubated for one week at 37°.

Cellulose Digestion (Mann, 1968)

The tubes are incubated for several weeks at 37°.

For use with the traditional anaerobic techniques

Unless otherwise stated all agar plates are stored for several days before use in an anaerobic jar containing a mixture of hydrogen and 10% carbon dioxide together with a catalyst to ensure the rapid removal of the last traces of oxygen. After inoculation the plates are returned immediately to the anaerobic jar and incubated for one week at 37°.

BGP Agar Supplemented with Liver and Faecal Extract (BGPhlf)

The medium contains (g or ml litre^{-1}): tryptone (Oxoid) 10 g; Lab-Lemco (Oxoid) powder 2·4 g; yeast extract (Difco) 5 g; cysteine HCl 0·4 g; glucose 1 g; NaCl 5 g; Na_2HPO_4 4 g; liver extract 50 ml; chicken faecal extract 50 ml; haemin (40 μg ml^{-1}) 25 ml; agar (New Zealand) 12g, pH 7·2–7·4.

Liver Extract

Difco-dehydrated liver (27 g) is dissolved in 200 ml distilled water, heated to 50° and held at this temperature for 1 h. The mixture is then boiled, cooled and centrifuged. The pH is adjusted to 7·0–7·2 and the supernatant sterilized by autoclaving at 121° for 15 min.

Chicken Faecal Extract

This is prepared by autoclaving at 121° for 30 min equal quantities of chicken faeces and water. The sludge is centrifuged and the supernatant poured off and adjusted to pH 7·0–7·2. It is sterilized by autoclaving at 121° for 15 min.

Supplemented Uric Acid Agar (Barnes and Impey, 1974)

Ethyl Violet Azide Agar

This is RCM agar (Hirsch and Grinsted, 1954) containing ethyl violet 1/20 000 and sodium azide 1/20 000. (Barnes and Goldberg, 1962.)

Bifid Agar

This is a modification of the medium of Mitsuoka *et al.* (1965) described by Barnes *et al.* (1972).

China Blue Agar

This is modified from van der Wiel-Korstanje and Winkler (1970). The medium consists of BGPhlf agar without the Na_2HPO_4 with the glucose increased to 10 g litre^{-1} and containing china blue (3%) 10 ml litre^{-1}. The china blue must be the ferri-ferrocyanide complex (No. 10365, Chroma-Gesellschaft: Stuttgart). The bifidobacteria appear as brown colonies.

Sulphite Polymyxin-Sulphadiazine Medium (SPS) of Angelotti et al. *(1962)*

The deep agar tubes are incubated in air for two days at 37°. Clostridia appear as black colonies.

Willis and Hobbs Agar (Willis and Hobbs, 1958)

The medium is modified by omitting the milk and neutral red indicator but incorporating 100 μg ml^{-1} of neomycin sulphate. Plates are stored aerobically and after inoculation are incubated for 18–24 h at 37° in an anaerobic jar under hydrogen. Colonies surrounded by a zone of opalescence in the medium are picked for further confirmatory tests for *Clostridium perfringens* (see below).

For the facultative anaerobes

Rogosa Agar (Oxoid) for Lactobacilli

The plates are incubated at 37° for two days in an atmosphere of hydrogen and 10% carbon dioxide.

Thallous Acetate Tetrazolium Agar (TlTG) Barnes (1956)

The constituents of the medium in g litre^{-1} are: peptone (Evans) 10; Lab-Lemco powder (Oxoid) 8; glucose 10; thallous acetate 1; triphenyltetrazolium chloride 0·1; agar (New Zealand) 14, pH 6·0. The thallous acetate (5% autoclaved solution) is added to the autoclaved medium (20 ml litre^{-1}) together with 10 ml litre^{-1} of a 1% solution of tetrazolium sterilized by membrane filtration. The plates are incubated for one day at 37° and then left for several days at room temperature. On this medium colonies of *Streptococcus faecalis* and subspecies *liquefaciens* and *zymogenes* appear as red or red-centred whilst *Strep. faecium* and related strains are white or pale pink.

Tyrosine Agar (Mead, 1963, 1964)

The medium contains (g litre^{-1}): peptone (Evans) 10; yeast extract (Oxoid) 1; sorbitol 2; tyrosine 5; thallous acetate 1; triphenyltetrazolium chloride 0·1; agar (New Zealand) 12, pH 6·0–6·2. Solutions of thallous acetate and triphenyltetrazolium chloride are prepared and added to the medium as for T1TG agar. Layered plates are prepared with the above medium as the lower layer. For the upper layer a further 1·5 g 100 ml^{-1} of tyrosine is added to the molten basal medium to form a milky suspension. Plates are incubated for three days at 37°. The medium is specific for certain biotypes of *Strep. faecalis*, subspecies *liquefaciens* and *zymogenes*. Typical colonies are dark red or red-centred and surrounded by a zone of clearing in the medium due to tyrosine decarboxylase activity.

MacConkey No. 3 Agar (Oxoid)

This is for the isolation of *coli-aerogenes* bacteria and other enterobacteria particularly *Proteus* spp. The plates are incubated at 37° for one day.

Media and method for salmonellae

The WHO recommended procedure described by Edel and Kampelmacher (1969) is used.

Isolation Procedures

The method generally used for the analysis of the intestinal flora is outlined below; details of media and incubation are given above. The justification for handling the sample in this particular way and the types of media used are discussed in the following sections concerned with the individual groups of organisms.

Preparation and dilution of the sample

With any sample which may contain anaerobic bacteria it is essential to take strict precautions to prevent contact with oxygen. Thus about 1 g of the sample is taken directly into a weighed tube containing 9 ml of anaerobic dilution solution (ADS) and some glass beads, this and subsequent operations being carried out under a continuous flow of oxygen-free carbon dioxide following the technique of Hungate (1950, 1969). After weighing and shaking, the suspension is used as follows.

(1) For the direct microscopical count (see p. 94).

(2) For the isolation of the obligate anaerobes using the Hungate technique (Method 1, below) when tenfold dilutions are prepared in ADS, generally to 10^{-9}. One millilitre of each of the required dilutions is added in duplicate or triplicate to 9 ml of pre-reduced molten SM10 agar in stoppered tubes and roll-tubes are prepared. The same dilutions are also used for plating on various pre-reduced differential or selective agar plates by rapidly spreading 2 drops (1/15 ml) of the required dilution over the surface of half plates which are then incubated in an anaerobic jar under hydrogen + 10% carbon dioxide (Method 2, below).

(3) For the isolation of certain clostridia and the facultative anaerobes 1 ml is taken and serially diluted in 9 ml quantities of Reinforced Clostridial Medium (RCM, Oxoid) from which the oxygen has been expelled by holding in a boiling water bath for 20 min immediately before use. These dilutions are then used in the following way (a) for sulphite-reducing clostridia—1 ml inocula of appropriate dilutions are added to sterile plastic-capped test tubes followed by freshly prepared molten SPS agar held at 50–52°. The agar is solidified rapidly by immersing the tubes in

cold water and then sealing the surface with a layer of uninoculated medium to a depth of about 2 cm. The tubes are incubated in air. (b) For the isolation of *Cl. perfringens* 1–3 drops (each of 1/30 ml) are spread over the surface of modified Willis and Hobbs agar. The plates are incubated in an anaerobic jar under hydrogen. (c) For the isolation of the facultative anaerobes, one drop (1/30 ml) of the appropriate dilution is spread over one-quarter of a Petri dish containing the required agar medium. The media used, together with the incubation times, are listed above.

Enrichment for pathogens

Where organisms such as salmonellae are expected to be present in very low numbers, part of the intestinal sample may be taken directly into the appropriate pre-enrichment or enrichment medium (see p. 93).

Direct microscopical count

It is essential to perform a direct microscopical count on every sample. It is only by doing this that a reference point is obtained to assess the efficiency of the isolation procedures used. The suspension in the anaerobic dilution solution (ADS) is diluted to the required concentration in saline (0·9% w/v) containing formalin (10%). The organisms are counted in a Helber counting chamber using the phase contrast microscope and following the method described by Meynell and Meynell (1965). Whether individual cells or colony forming units are counted will make a considerable difference to any figures obtained for the proportion of organisms actually isolated.

Isolation of Anaerobes

Both non-sporing anaerobes and clostridia vary considerably in their sensitivity to oxygen and may be grown using one of the following approaches.

Method 1

The Hungate technique for strict anaerobes (Hungate, 1950, 1969). Media preparation, dilution and incubation are carried out in the presence of an oxygen-free gas, usually 100% CO_2.

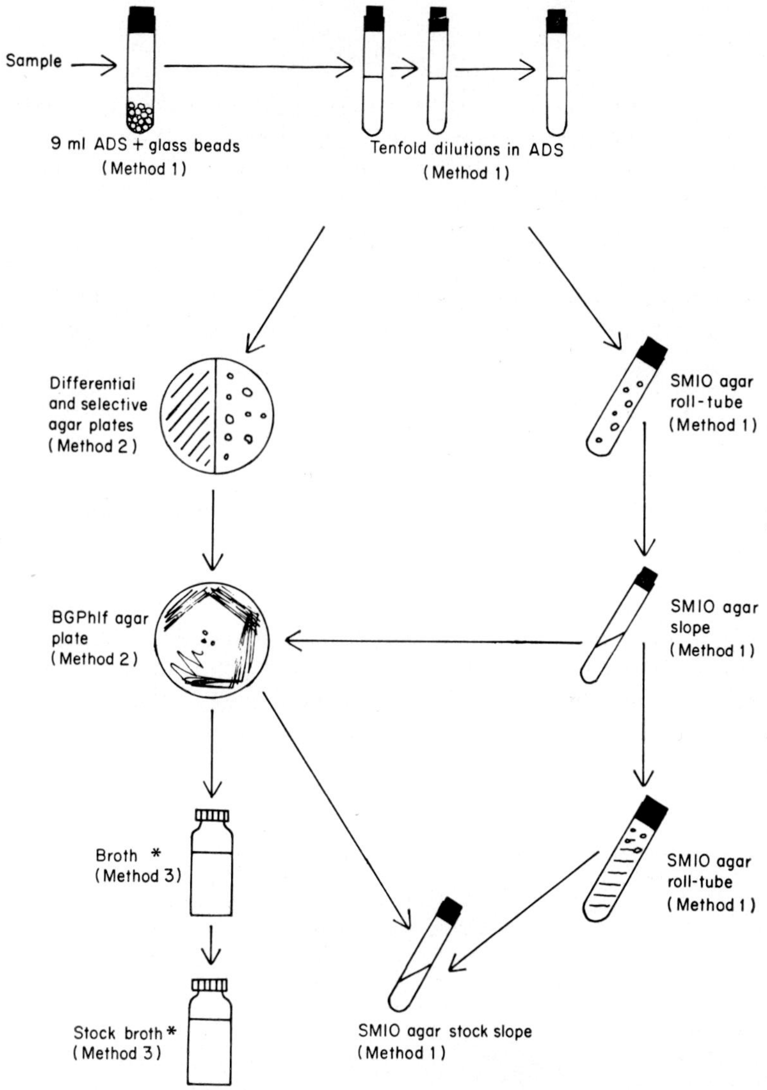

FIG. 1. General scheme for the isolation and purification of the anaerobic flora. * Data of Barnes and Impey (1968).

Method 2

Inoculation of pre-reduced agar plates and incubation in an anaerobic jar containing 90% hydrogen + 10% CO_2 together with a catalyst to remove traces of oxygen.

Method 3

Growth in media containing reducing agents either as deep agar in tubes, or broths in sealed or capped bottles. Any dissolved oxygen is removed immediately before inoculation by holding the media in a boiling water bath for 20 min.

Sufficient evidence has been obtained from tests with the intestinal anaerobes to show that Method 1 must be used for the primary isolation of certain organisms. Once an organism is isolated every effort is then made to use the simpler methods, following the scheme outlined in Fig. 1.

Non-sporing anaerobes

The ability of the non-sporing anaerobes to grow by Methods 1–3 given above, was tested using 37 types isolated freshly from the avian caecum. The results given in Table 1 indicate that without the use of the Hungate technique (Method 1) most of the recently described anaerobic budding bacteria (Gossling and Moore, 1975), which were found by Barnes *et al.* (1972) to form more than 13% of the total population of chickens aged two to six weeks, could not have been isolated. The growth of a number of other strains which are still being characterized is also dependent on the use of this technique. It is evident from Table 1 that many more

TABLE 1. The effect of the anaerobic technique used on the growth of some of the non-sporing anaerobic bacteria

Genus	Nos of species tested	Nos of species able to grow: Method 1 Using the Hungate technique	Method 2 On plates in anaerobic jar	Method 3 In broth in 1 oz screw-capped bottles
Bacteroides	9	9	9	8
Fusobacterium	3	3	3	2
Streptococcus (anaerobic spp.)	3	3	3	3
Peptostreptococcus	7	7	7	6
Eubacterium	9	9	9	6
Bifidobacterium	1	1	1	1
Budding bacteria*	5	5	2	0
Total	37	37	34	26

* The first of the anaerobic budding bacteria has now been assigned to the genus *Gemmiger* by Gossling and Moore (1975).

organisms can grow on pre-reduced agar plates incubated in an anaerobic jar under hydrogen and carbon dioxide 10% (Method 2) than in a reducing medium in screw-capped bottles (Method 3).

The most satisfactory medium for isolating and maintaining stock cultures has been found to be Medium 10 supplemented with liver and chicken faecal extract (SM10). Medium 10 was developed originally by Caldwell and Bryant (1966) for isolating the rumen flora. However, preliminary trials showed that only a very small proportion of anaerobes from chicken caeca could be isolated with this medium. The medium was then supplemented with chicken faecal extract and liver extract (SM10) and a high recovery of organisms was obtained. Many of the organisms which were isolated using SM10 agar have now been tested for their ability to grow in the unsupplemented Medium 10. It can be seen from Table 2 that whilst 9 out of 13 of the Gram-negative anaerobic rods (*Bacteroides* and *Fusobacterium*) could grow in the unsupplemented M10, almost all of the Gram-positive cocci and rods required faecal extract which in some cases could be replaced by liver extract. Two

TABLE 2. The ability of the non-sporing anaerobic bacteria to grow (using the Hungate technique) in Medium 10 with or without supplements of chicken faecal extract (10%) and liver extract (5%)

Organism		Optimal growth in M10 alone	M10 with faecal extract	M10 with liver extract
Bacteroides	*fragilis*	+	nt†	nt
	hypermegas	+	nt	nt
	clostridiiformis var. *clostridiiformis* EBF 77/78‡	—	+	—
	sp. EBF 59/96*‡	+	nt	nt
	sp. NE3/203‡	—	+	—
	sp. NE1/8‡	+	nt	nt
	sp. EBF 77/26A	+	nt	nt
	sp. NE3/254	—	+	—
	sp. NE1/26	+	nt	nt
Fusobacterium	*necrogenes*	+	nt	nt
	plauti NE3/244‡	—	+	—
	NE3/253‡	+	nt	nt
	sp. NE1/74	+	nt	nt
Streptococcus	*intermedius*	—	+	—
	sp. EBF 61/60B†	—	+	—
	sp. EBF 77/14B‡	—	+	—
Coprococcus sp. NE1/97§		—	+	—

Table 2—continued

Organism		Optimal growth in M10 alone	M10 with faecal extract	M10 with liver extract
Peptostreptococcus	sp. NE3/225 group 2 coccus†	—	+	—
	sp. NE3/239‡	+	nt	nt
	sp. NE3/199‡	—	+	+
	sp. NE1/71‡	—	+	+
	sp. NE3/250‡	—	+	+
	sp. NE1/51A	—	+	—
Eubacterium	sp. NE3/191‡	—	+	+
	sp. NE3/198	—	+	—
	sp. NE3/197	+	nt	nt
	sp. EBG1/80	—	+	—
	sp. EBF77/35	—	+	—
	sp. EBF77/24A	—	+	—
	sp. EBG1/141	—	+	—
	sp. NE3/230B	+	nt	nt
	sp. NE3/98	—	+	+
Bifidobacterium sp. EBF77/14A		—	+	—
Curved rod (Gram + ve) NE1/22		—	—	+
Budding bacteria (anaerobic Gram + ve or — ve)				
Gemmiger formicilis	NE3/247	—	—	+
	NE3/265	—	+	—
	NE3/217	—	+	—
	NE3/235	—	+	—
	NE3/209	+	nt	nt

* Barnes and Impey (1968); † 1970; ‡ 1974; § Holdeman and Moore (1974); nt, not tested. All of the organisms grew well in SM10 broth.

organisms—an unidentified curved Gram-positive rod and one of the budding bacteria, *Gemmiger formicilis*, required liver extract and failed to grow with faecal extract alone. Analyses of the faecal extract have been made (Barnes and Impey, 1974) but the particular substances required by the individual species is not known.

After growth the colonies in the anaerobic roll-tubes are purified as shown in Fig. 1. They are first picked into SM10 agar slopes and after 24 h incubation or as soon as growth has occurred, they are examined microscopically; the cultures are also tested for aerobic growth in order to eliminate any facultative anaerobes which may have been isolated. The SM10 agar culture is then streaked on to pre-reduced BGPhlf agar plates which are incubated in the anaerobic jar under hydrogen and 10% carbon dioxide. If growth occurs, an isolated colony is transferred into a new SM10 agar slope for stock purposes. If no growth has occurred on the plate, purification and all further testing has to be carried out using the Hungate technique. Details of the tests used to identify the organisms have been described by Barnes and Impey (1974).

By following the above procedure about 40 types of non-sporing anaerobic bacteria have been isolated so far from the caeca, occurring at $> 10^9\ g^{-1}$. The majority of these are listed in Table 2 but many can only be identified as far as the genus. In addition several strict anaerobes have been isolated which have still not been characterized.

Use of selective and differential media

For the isolation of anaerobes with properties of particular interest selective media may be used, either in combination with Method 1, e.g. as for cellulolytic organisms (Mann, 1968) or more easily for the less exacting anaerobes by using pre-reduced agar plates and the traditional anaerobic jar technique (Method 2). Using a multipoint inoculating technique Barnes and Impey (1972) tested the ability of a number of the anaerobic bacteria to attack various substrates of possible ecological significance (starches, proteins, nucleic acids, etc.) as well as their behaviour in the presence of some inhibitors. This led to the development of a number of selective and differential media, one of the most interesting being for the detection of uric acid-decomposing organisms (Barnes and Impey, 1974).

At present, the number of selective media for isolating particular groups of anaerobes is very limited. *Bacteroides hypermegas* can be isolated using an ethyl violet azide medium (see p. 91) whilst *Bifidobacterium* spp. are isolated either on Bifid agar or on China Blue agar (see p. 91) but neither of these media have been found to be highly selective. Finegold, Sugihara and Sutter (1971) have developed a number of selective media for the isolation of anaerobes from humans but they are not always applicable to the intestinal flora of the bird as the types of organisms present tend to be different.

Clostridia

Many different clostridia can be isolated from the avian intestine and the organisms are sometimes present among the predominant flora of the caeca. Most of the known species can be grown by using any of the techniques described in this paper for the less demanding anaerobes including the use of liquid media in screw-capped bottles, deep agar tubes and anaerobic plates (Methods 2 and 3). However, some strains from the predominant flora of the chicken caecum which are not identifiable with known species have failed to grow when tested in reducing media contained in screw-capped bottles (Method 3). Growth of these strains was obtained on anaerobic plates only when particular care was taken to minimize contact with air. Species isolated from chickens, turkeys and pheasants are shown in Table 3. Many other strains which cannot be identified have also been isolated.

TABLE 3. Species of *Clostridium* isolated from intestinal contents of chickens, turkeys and pheasants

Species	Source
Cl. bifermentans	c, p
Cl. carnis	c
Cl. histolyticum	c
Cl. innocuum	c
Cl. lentoputrescens	c
Cl. malenominatum	c
Cl. paraperfringens	c
Cl. paraputrificum	c, t
*Cl. perenne**†	p
Cl. perfringens	c, t, p
Cl. putrefaciens†	p
Cl. sartagoformum	c
Cl. sordellii	p
Cl. sporogenes	c, t
Cl. subterminale	c
Cl. symbiosum‡	c
Cl. tertium	c
Cl. tetanomorphum	t
Cl. tyrobutyricum	c

c, chickens; t, turkeys; p, pheasants.
* Previously unidentified strains (Mead and Chamberlain, 1971) identified by L. D. S. Smith (pers. comm.).
† Isolated only from hung birds.
‡ Kaneuchi *et al.* (1976).

Most clostridia can be isolated following dilution of the sample in RCM. No selective medium is available which will permit the isolation of all types present and for the predominant strains the best approach is to pick colonies from non-selective media and examine them for spore formation. It is often useful to heat samples at 70° for 10 min before culturing in order to destroy non-sporing organisms; some strains of clostridia from the avian intestine produce spores which fail to survive heating at higher temperatures.

Sulphite-reducing clostridia

Many of the clostridia present are able to grow and produce black colonies in media containing sulphite and ferric ions by virtue of their ability to reduce the sulphite with the subsequent precipitation of iron sulphide. The medium used for their isolation is the sulphite-polymyxin-sulphadiazine (SPS) medium of Angelotti *et al.* (1962) (see p. 91). The main disadvantage of the medium is that some intestinal bacteria other than clostridia may produce black colonies and hence it is necessary to pick colonies from cultures prepared from the highest dilution of the sample to confirm the presence of spore-forming organisms. Another disadvantage is that SPS medium is unsuitable for samples containing high numbers of faecal streptococci which grow well in the medium and interfere with the growth of clostridia. This problem has been encountered with caecal samples from young chicks in which up to 10^{10} faecal streptococci g^{-1} are present.

Clostridium perfringens

The organism is important in human food poisoning, avian disease and possibly as an agent in avian growth depression. Because it is less sensitive to oxygen than most anaerobic bacteria, this property can be exploited to reduce interference from other anaerobes on selective media and is particularly useful in dealing with caecal samples which contain a variety of interfering organisms.

Counts are obtained by surface-plating on the modified Willis and Hobbs egg yolk medium (see p. 91) which is deliberately rendered less suitable for many anaerobes by (a) holding plates aerobically at room temperature for one week before use and (b) incubating plates under H_2 rather than a gas mixture containing CO_2 which is required for growth by many other caecal anaerobes.

An egg yolk medium is preferred for selective isolation because few other anaerobic organisms found in the avian intestine produce egg yolk reactions (Barnes and Impey, 1972). However, several species of clostridia produce a zone of opalescence in the medium so confirmatory tests must be carried out. The organisms are picked for further tests using the same medium plus neutral red: these are (a) inhibition of the egg yolk reaction on "half antitoxin" plates containing two drops (1/15ml) of *Cl. perfringens* type A antitoxin spread over the surface; (b) fermentation of lactose, following anaerobic incubation, plates are left on the bench for a day to allow the pH indicator to become reoxidized. A test for gelatin liquefaction is also included. These latter tests differentiate *Cl. perfringens* from *Cl. bifermentans*, *Cl. paraperfringens*, lecithin-positive *Cl. perenne* and *Cl. sordellii* which also produce a lecithinase susceptible to inhibition by *Cl. perfringens* type A antitoxin. Among the five species only *Cl. perfringens* ferments lactose and liquefies gelatin.

Isolation of the Facultative Anaerobes

The most important groups of facultative anaerobes in the intestines are the lactobacilli, faecal streptococci and enterobacteria. Their numbers vary considerably both with the age of the bird and the region of the intestine. It has been shown that quite large differences in diet or the use of feed additives may not alter the numbers of organisms present but there may be significant changes in the type of organism found. Thus the use of media which can detect strain or species differences may be important.

Total aerobic count

The precaution is taken of carrying out a total aerobic count using a rich medium such as Bacto Brain Heart Infusion agar (Difco) or MRS agar (Oxoid) in order to recover any organisms present in large numbers which may not be isolated on the selective media.

Lactobacilli

The main types of lactobacilli present are reported to be *Lactobacillus acidophilus*, *L. salivarius* and *L. fermenti* (Morishita *et al.*, 1971). It is uncertain whether there are any additional types in the intestine which cannot grow on the Rogosa agar used for the isolation of lactobacilli in general.

Faecal streptococci

Streptococcus faecalis, subspecies *liquefaciens* and *zymogenes* together with atypical strains of *Strep. faecium* are usually found. Another species, *Strep. avium*, has been reported by Nowland and Deibel (1967). We have recently isolated unidentified types which sometimes outnumber all the other streptococci present; these are being studied. It has been found that although the numbers of faecal streptococci may remain constant the type may vary especially in the young chick (unpublished). The significance of such changes needs further study particularly in relation to the incidence of *Strep. faecalis* subspecies *zymogenes* and *liquefaciens*. For this reason two media are frequently used for faecal streptococci—

one, T1TG agar (see p. 92) will differentiate several types of streptococci, the other Tyrosine agar (see p. 92) selects only particular types of *Strep. faecalis* and subspecies. The difficulty with the T1TG medium is that lactobacilli tend to grow and may be confused with very small colonies of streptococci after 24 h incubation at 37°. It has been found that by leaving the incubated plates at room temperature for several days, the streptococcal colonies increase in size and thus can be differentiated more easily.

Enterobacteria

The *coli-aerogenes* bacteria are generally isolated on MacConkey No. 3 agar (Oxoid) and *Proteus* spp. when present will also be detected on this medium. The salmonellae occur only in very low numbers if at all in the healthy bird and special techniques are used for their recovery (see p. 93)

Discussion

In the analysis of the microflora of the intestine one is dealing with a very complex situation where possibly between 50 and 100 different types of organisms are existing together at any one time. Studies comparing individual birds from the same group have shown that the composition of the flora is remarkably constant and when it changes as with the age of the bird it does so in a consistent manner. The ecological significance of this flora is still not understood. It may be that in the analysis of such an environment quite different types of selective media are needed to answer particular questions. It was only when a search was made for organisms which could decompose uric acid, the main nitrogenous excretory product of the bird, under anaerobic conditions, that certain *coli-aerogenes* bacteria and faecal streptococci were found to possess this property as well as the many different types of caecal anaerobes (Mead and Adams, 1975). The significance of this to the bird is still a matter of conjecture relating to the function of the caecum. Similarly, there may be other properties shared by the intestinal flora which have no relevance in ordinary taxonomic studies. Thus, the analysis of such an environment should not be just a matter of isolating quantitatively those organisms which belong to a particular genus or species but their properties should be examined in relation to their particular habitat.

References

ANGELOTTI, R., HALL, H. E., FOTER, M. J. & LEWIS, K. H. (1962). Quantitation of *Clostridium perfringens* in foods. *Appl. Microbiol.*, **10**, 193.

BARNES, E. M. (1956). Methods for the isolation of faecal streptococci (Lancefield group D) from bacon factories. *J. appl. Bact.*, **19**, 193.

BARNES, E. M. (1975). Development and ecological significance of the avian intestinal flora. In *Proceedings of the first intersectional congress of IAMS*, Vol. 2 (Hasegawn, T., ed.). Japan, Science Council of Japan, p. 307.

BARNES, E. M. & GOLDBERG, H. (1962). The isolation of Gram-negative bacteria from poultry reared with and without antibiotic supplements. *J. appl. Bact.*, **25**, 94.

BARNES, E. M. & IMPEY, C. S. (1968). Anaerobic Gram-negative non-sporing bacteria from the caeca of poultry. *J. appl. Bact.*, **31**, 530.

BARNES, E. M. & IMPEY, C. S. (1970). The isolation and properties of the predominant anaerobic bacteria in the caeca of chickens and turkeys. *Br. Poult. Sci.*, **11**, 467.

BARNES, E. M. & IMPEY, C. S. (1972). Some properties of the non-sporing anaerobes from poultry caeca. *J. appl. Bact.*, **35**, 241.

BARNES, E. M. & IMPEY, C. S. (1974). The occurrence and properties of uric acid decomposing anaerobic bacteria in the avian caecum. *J. appl. Bact.*, **37**, 393.

BARNES, E. M., MEAD, G. C., BARNUM, D. A. & HARRY, E. G. (1972). The intestinal flora of the chicken in the period 2–6 weeks of age with particular reference to the anaerobic bacteria. *Br. Poult. Sci.*, **13**, 311.

BRYANT, M. P. & BURKEY, L. A. (1953). Cultural methods and some characteristics of some of the more numerous groups of bacteria in the bovine rumen. *J. Dairy Sci.*, **36**, 205.

CALDWELL, D. R. & BRYANT, M. P. (1966). Medium without rumen fluid for nonselective enumeration and isolation of rumen bacteria. *Appl. Microbiol.*, **14**, 794.

EDEL, W. & KAMPELMACHER, E. H. (1969). *Salmonella* isolation in nine European laboratories using a standardised technique. *Bull. Wld Hlth Org.*, **41**, 297.

FINEGOLD, S. M., SUGIHARA, P. T. & SUTTER, V. L. (1971). Use of selective media for the isolation of anaerobes from humans. In *Isolation of anaerobes* (Shapton, D. A. & Board, R. G., eds). Soc. appl. Bact. Tech. Ser. No. 5. London and New York: Academic Press, p. 99.

GOSSLING, J. & MOORE, W. E. C. (1975). *Gemmiger formicilis* n. gen., n. sp., an anaerobic budding bacterium from intestines. *Int. J. Syst. Bact.*, **25**, 202.

HIRSCH, A. & GRINSTED, E. (1954). Methods for the growth and enumeration of anaerobic spore formers from cheese with observations on the effect of nisin. *J. Dairy Res.*, **21**, 101.

HOLDEMAN, L. V. & MOORE, W. E. C. (1974). New Genus, *Coprococcus*, twelve new species and emended descriptions of four previously described species of bacteria from human faeces. *Int. J. Syst. Bact.*, **24**, 260.

HUNGATE, R. E. (1950). The anaerobic mesophilic cellulolytic bacteria. *Bact. Rev.*, **14**, 1.

HUNGATE, R. E. (1969). A roll tube method for cultivation of strict anaerobes. In *Methods in microbiology*, Vol. 3B (Norris, J. R. and Ribbons, D. W., eds). London and New York: Academic Press, p. 117.

JAYNE-WILLIAMS, D. J. & FULLER, R. (1971). The influence of the intestinal microflora on nutrition. In *Physiology and biochemistry of the domestic fowl* (Bell, D. J. & Freeman, B. M., eds). New York and London: Academic Press, p. 73.

KANEUCHI, C., WATANABE, K., TARADA, A., BENNO, Y. & MITSUOKA, T. (1976). Taxonomic study of *Bacteroides clostridiiformis* subsp. *clostridiiformis* (Burri and Ankersmit) Holdeman and Moore and related organisms. Proposal of *Clostridium clostridiiformis* (Burri and Ankersmit) comb. nov. and *Clostridium symbiosum* (Stevens) comb. nov. *Int. J. Syst. Bact*, **26**, 195.

MANN, S. O. (1968). An improved method for determining cellulolytic activity in anaerobic bacteria. *J. appl. Bact.*, **31**, 241.

MEAD, G. C. (1963). A medium for the isolation of *Streptococcus faecalis* sensu strictu. *Nature, Lond.*, **197**, 1323.

MEAD, G. C. (1964). Isolation and significance of *Streptococcus faecalis* sensu strictu. *Nature, Lond.*, **204**, 1224.

MEAD, G. C. & ADAMS, B. W. (1975). Some observations on the caecal microflora of the chick during the first two weeks of life. *Br. Poult. Sci.*, **16**, 169.

MEAD, G. C. & CHAMBERLAIN, A. M. (1971). An unusual species of *Clostridium* isolated from the intestine of the pheasant. *J. appl. Bact.*, **34**, 815.

MEYNELL, G. C. & MEYNELL, E. (1965). *Theory and practice in experimental bacteriology*. Cambridge: Cambridge Univ. Press, p. 12.

MITSUOKA, T., SEGA, T. & YAMAMOTO, S. (1965). Eine verbesserte Methodik der qualitativen und quantitativen Analyse der Darmflora Von Menschen und Tieren. *Zentbl. Bakt. Parasitkde* (*Abt. I*), **195**, 455.

MORISHITA, Y., MITSUOKA, T., KANEUCHI, C., YAMAMOTO, S. & OGATA, M. (1971). Specific establishment of lactobacilli in the digestive tract of germ-free chickens. *Japan J. Microbiol.*, **15**, 531.

NOWLAND, S. S. & DEIBEL, R. H. (1967). Group Q streptococci. 1. Ecology, serology, physiology and relationship to established enterococci. *J. Bact.*, **94**, 291.

SHRIMPTON, D. H. (1966). Metabolism of the intestinal microflora in birds and its possible influence on the composition of flavour precursors in their muscles. *J. appl. Bact.*, **29**, 222.

VAN DER WIEL-KORSTANJE, J. A. A. & WINKLER, K. C. (1970). Medium for differential count of the anaerobic flora in human faeces. *Appl. Microbiol.*, **20**, 168.

WILLIS, A. T. & HOBBS, G. (1958). A medium for the identification of clostridia producing opalescence in egg-yolk emulsions. *J. Path. Bact.*, **75**, 299.

The Isolation and Use of Streptomycin-resistant Mutants for Following Development of Bacteria in Mixed Cultures

R. W. A. Park

University of Reading, Reading, Berkshire, England

Introduction

When one wishes to study the changes in numbers of a particular type of bacterium in a habitat containing many types of bacteria the problems are considerable. Selective enumeration media have been devised for some bacteria, but the degree of selectivity exerted is seldom sufficient to allow enumeration of any one type in the presence of closely related types. Of course, for the great majority of bacteria no selective media are available. Several studies of the development of one kind of bacterium in a complex habitat have involved use of antibiotic-resistant mutants. These mutants could be enumerated in the presence of many other bacteria by incorporation of the appropriate antibiotic into a growth medium for plate counts (Greenberg, 1969; Obaton, 1971, 1973; Danso *et al.*, 1973; Akman and Park, 1974). Mutants resistant to large concentrations of streptomycin seem to have been most useful. Four aspects of the use of streptomycin-resistant mutants for studying the population dynamics of one type of bacterium in the presence of other types are considered here.

Obtaining a Streptomycin-resistant Mutant

Bacteria occasionally mutate to become resistant to 1000 μg ml^{-1} streptomycin. The aim of this technique, described by Meynell and Meynell (1970), is to allow such a mutant to arise and be isolated. The procedure is as follows.

(1) Inoculate the wild type into a large amount of suitable liquid growth medium (e.g. 100 ml nutrient broth in a 250 ml conical flask).

(2) Incubate until the culture is turbid (e.g. overnight, shaken at 37° if enterobacteria). At this stage most of the population will be

sensitive to streptomycin but a few resistant mutants are likely to be present, having developed spontaneously.

(3) Add an equal volume of growth medium containing 2000 μg ml^{-1} streptomycin and reincubate. Thus the concentration of antibiotic in the medium is 1000 μg ml^{-1}, the wild type will be inhibited and any resistant mutants present will grow to a large population using the nutrients just provided.

(4) After incubation (e.g. enterobacteria 8 h and 24 h) remove 0·1 ml culture and spread across a plate of solid growth medium containing 1000 μg ml^{-1} streptomycin. Use a loop to remove some of this inoculum to streak on to similar plates to obtain isolated colonies.

(5) After incubation restreak from an isolated colony on to solid medium containing 1000 μg ml^{-1} streptomycin and incubate.

(6) Use an isolated colony from the medium inoculated in (5) to establish a stock of culture on an appropriate growth medium not containing streptomycin.

(7) Periodically the culture should be checked, by comparing counts on media with and without streptomycin, to determine whether any reversion to sensitivity is occurring. Stock cultures should be maintained free from streptomycin to avoid development of dependence.

There seems to be no reason *a priori* for thinking that this technique would work with only a few types of bacteria. There are many reports of mutants resistant to 1000 μg ml^{-1} streptomycin and we have obtained in this laboratory such mutants from *Escherichia coli*, *Lactobacillus arabinosus*, *Pseudomonas fluorescens*, various *Salmonella* spp., *Staphylococcus aureus* and non-fermenting vibrios. A similar principle but using other antibiotics may be applied to the study of growth of yeasts and moulds.

Examples of the Use of Mutants to Study Population Dynamics in Complex Environments

The growth of Salmonella *spp. on cured meats*

Streptomycin-resistant mutants of various *Salmonella* spp. were added in known numbers to separate small pieces of the meats. The rate of development of the salmonellas was assessed by periodically mascerating one of the pieces and plating out appropriate dilutions on nutrient agar containing 1000 μg ml^{-1} streptomycin. Differences between meats in their ability to support growth of salmonellas, and the effects of various storage temperatures, were readily detected (Akman and Park, 1974). The use of mutants avoided complications that might otherwise arise from uneven distribution of other salmonella on the meat as supplied and from any lack of "definition" of the usual selective media.

Comparisons between Salmonella *and* Escherichia coli *with respect to survival in samples of river water*

It is often thought that *E. coli* survives longer in the natural environment than do enteric pathogens, though there appears to be little evidence on this point. The use of streptomycin-resistant mutants of *Salmonella* and *E. coli* facilitated a comparison of survival in separate samples of unsterilized water from a river (Fig. 1).

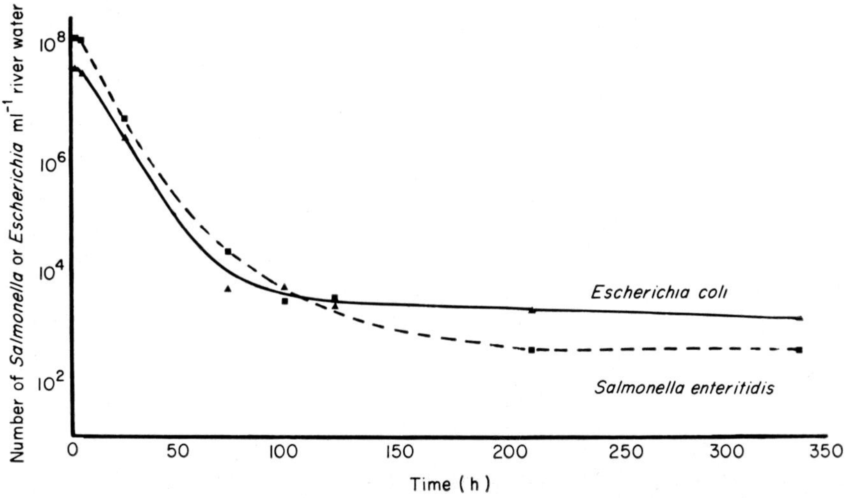

FIG. 1. Comparison of survival of *Escherichia coli* and *Salmonella enteritidis* at 17° in river water containing the normal microbial flora. ▲, *Escherichia coli*; ■, *Salmonella enteritidis*.

Overnight cultures in nutrient broth at 37° of a streptomycin-resistant mutant derived from *S. enteritidis* (Salmonella Reference Laboratory 4753/70) and a streptomycin-resistant mutant derived from *E. coli*. (NCTC 9434) were inoculated (0·5 ml) into separate 500 ml conical flasks containing 300 ml freshly collected river water. The inocula were dispersed by shaking and the flasks were then incubated, static, at 17°. Samples were removed periodically, and diluted and counted by the Miles and Misra Method using nutrient agar + 1000 μg ml^{-1} streptomycin.

Is an organism isolated from an environment capable of growth in that environment?

Jangi (1975) isolated 30 non-fermentative vibrios from various surface waters around Reading. Such organisms appear to be common in the waters, but are they capable of growth in these habitats or do they enter from other habitats and not grow in the waters? Addition of a streptomycin-resistant mutant of one of these isolates to a sample of river water, and periodic enumeration using streptomycin containing agar, has indicated that it is unlikely that the isolate was active in water (Fig. 2) in large numbers.

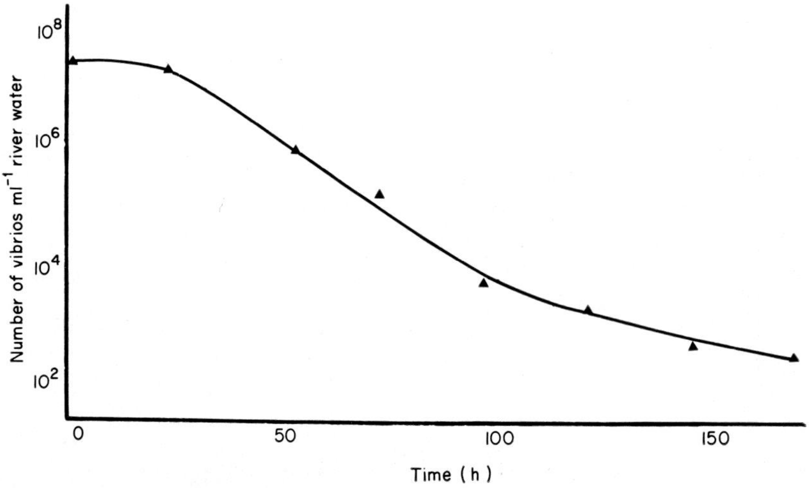

FIG. 2. Survival at 18° of a non-fermenting vibrio in river water containing the normal microbial flora

A streptomycin-resistant mutant derived from a non-fermenting vibrio (Dr. Jangi strain 14) was incubated for 24 h at 30° in nutrient broth + 0·3% yeast extract (Oxoid). The culture was then centrifuged, re-suspended to original volume in Ringer's solution and 0·4 ml added to 400 ml freshly collected river water held in a 500 ml conical flask and stirred with a magnetic stirrer. Incubation was at 18° and samples were removed periodically and counted by the Miles and Misra Method, using nutrient agar + 0·3% yeast extract + 1000 μg ml^{-1} streptomycin.

Application of Findings with Mutants to the Wild Type

Relative growth rates of mutants and wild type

In using resistant mutants to study survival and growth one is making the assumption that the mutants and the wild type behave similarly. Such an assumption is not necessarily justified because streptomycin-resistant mutants are produced naturally and yet do not dominate the sensitive wild type in the absence of antibiotic. Two of the ways in which the wild type and the resistant mutant can be compared are (a) by comparing the rate of increase in turbidity of separate cultures of the wild type and mutant and (b) by counting during growth resistant bacteria and total bacteria in a culture inoculated with equal numbers of mutant and wild type.

Each method has revealed only a slightly slower growth rate of the resistant mutant or no difference and it seems that differences in growth rate between mutant and wild type are unimportant for most of the likely uses of the technique. For example, only slight or no difference was found between the growth rate of a streptomycin-resistant mutant of *Staphylococcus aureus* and the sensitive parent under a wide range of conditions in shaken nutrient broth (NaCl, 0·5 to 12·0%; initial pH 4·5 to 10·1; temperature 25 or 33°). However, Bennett and Billing (1975) and Russell (1975) have found that some plant pathogens lose their pathogenicity on becoming resistant to streptomycin.

Transfer of resistance to other strains

Resistance to 1000 $\mu g\ ml^{-1}$ streptomycin is conferred by modification of a chromosomal gene. Resistance to this large concentration is not carried on a plasmid. I have no evidence of transfer affecting the results of experiments involving resistant mutants.

Development of streptomycin dependence

Mutants can be isolated which can grow only in the presence of streptomycin. Clearly, such mutants would be of no use for following growth in complex habitats. To avoid development of these dependent mutants stock cultures of the resistant mutants are kept on media lacking streptomycin. This requires periodic examination of the stocks to see that reversion to the wild type is not occurring.

Other streptomycin-resistant bacteria in the environment

Numbers of bacteria in nature resistant to 1000 $\mu g\ ml^{-1}$ streptomycin are relatively small but the possibility of interference with experiments should be assessed by using appropriate controls.

Very large inocula may disturb the normal situation

One should bear in mind that inoculation of a large number of test organisms into a system may so change the conditions as to give results different from those which would be obtained with only a small number.

Use of the Mutants and the Ashby Report

Mutants resistant to large concentrations of streptomycin arise naturally. The procedure for their isolation described here involves only the application of a selective pressure (presence of the antibiotic) to separate them from the wild type. No genetic manipulation as described in the Ashby Report (Report, 1975) is involved. Nevertheless, it is prudent not to release resistant mutants into the environment, restricting their use to studies of laboratory simulations so that the organisms can be destroyed at the end of experiments.

Acknowledgement

I wish to thank Miss Susan Giles for technical assistance.

References

AKMAN, M. & PARK, R. W. A. (1974). The growth of salmonellas on cooked cured pork. *J. Hyg., Camb.*, **72**, 369.

BENNETT, R. A. & BILLING, E. (1975). Development and properties of streptomycin-resistant cultures of *Erwinia amylovora* derived from English isolates. *J. appl. Bact.*, **39**, 307.

DANSO, S. K. A., HABTE, M. & ALEXANDER, M. (1973). Estimating the density of individual bacterial populations introduced into natural ecosystems. *Canad. J. Microbiol.*, **19**, 1450.

GREENBERG, B. (1969). Salmonella suppression by known populations of bacteria in flies. *J. Bact.*, **99**, 629.

JANGI, M. S. (1975). *Vibrios and Spirilla from surface waters*. Ph.D. Thesis, Univ. of Reading.

MEYNELL, G. G. & MEYNELL, E. (1970). *Theory and practice in experimental bacteriology*, 2nd edn. Cambridge: Cambridge Univ. Press.

OBATON, M. (1971). Utilisation de mutants spontanes resistants aux antibiotiques pour l'atude ecologique des *Rhizobium*. *C.R. Acad. Sci., Paris,* **272**, 2630.

OBATON, M. (1973). The use of spontaneous antibiotic resistant mutants for the ecological study of *Rhizobium*. *Bull. Ecol. Res. Comm. Stockholm*. **17**, 170.

REPORT (1975). Report of the working party on the experimental manipulation of the genetic composition of micro-organisms. Cmnd. 5880. London: HMSO.

RUSSELL, P. E. (1975). Variation in the virulence of some streptomycin-resistant mutants of *Pseudomonas phaseolicola*. *J. appl. Bact.*, **39**, 175.

Use of a Serum Bottle Technique to Study Interactions Between Strict Anaerobes in Mixed Culture

M. J. LATHAM

National Institute for Research in Dairying, Shinfield, Reading, Berkshire, England

AND

M. J. WOLIN

Environmental Health Centre, New York State Department of Health, Albany, New York, USA

Introduction

The majority of natural or artificial anaerobic ecosystems are complex and contain many different species of bacteria. The numerically important species in the anaerobic ecosystem of the rumen have now been isolated in pure culture using techniques based on those first described by Hungate in 1950. The Hungate technique has been adopted not only by rumen microbiologists (Kistner, 1960; Bryant and Robinson, 1961) but also by those seeking to isolate and identify anaerobes from a wide variety of clinical and other sources (Holdeman and Moore, 1972; Mah and Sussman, 1968; Hobson and Shaw, 1974). The role of the anaerobes in their natural ecosystem has largely been inferred from their properties when grown in pure culture but very little is known about the way in which their growth and activity is influenced by the other microorganisms in their environment. A modification of the Hungate technique by which organisms are cultured in serum bottles was recently described by Miller and Wolin (1974) and this modification has since been used to study interactions between strict anaerobes in mixed culture, particularly those involving the production and utilization of hydrogen. In this paper some of the advantages of the serum bottle method in the preparation of media, inoculation and sampling of cultures and its application in the study of interactions between cellulolytic and methanogenic rumen bacteria will be described.

Materials and Equipment

Oxygen-free gas and gas distribution

An essential feature of the Hungate technique is the use of oxygen-free (O_2-free) gas to displace any air from the culture medium headspace. A complete description of the construction and use of an electrically heated copper column for the removal of oxygen from commercial gases was given by Latham and Sharpe (1971). A more recent modification is the use of a domestic room light dimmer switch in place of the more expensive transformer originally described, which together with a contact thermometer (Electrical Thermometer Co. Ltd, Napier Place, Thetford, Norfolk, England) gives adequate control of column temperature.

The O_2-free gas emerging from the column is passed into a manifold made from 15 mm copper tubing with several outlets each fitted with flow regulating needle valves. The gas is passed from the valves through butyl rubber tubing to gassing jets. These are made from stainless steel and comprise a barrel (15 mm wide × 70 mm long) packed with glass wool, fitted with a hose connector for the butyl rubber tubing at one end and a narrow bore (1·0 mm i.d.) stainless steel gas outlet tube at the other end. The gas outlet tube is bent at right angles close to the barrel and is usually either 80 or 30 mm long depending upon whether the jet is to be used for gassing the medium or its headspace. The complete gassing jet assembly is sterilized before use.

Serum bottles and fittings

Serum bottles ranging in capacity from 6 to 125 ml, seals made of several types of synthetic rubber, and aluminium caps, can be obtained from Pierce and Warriner (UK) Ltd, 44, Upper Northgate Street, Chester, Cheshire CH1 4EF or Wheaton Scientific, Millville, New Jersey, USA. All bottles have the same sized neck so that seals and caps are interchangeable. Bottles of 15 ml capacity are most suitable for routine culture of organisms. Butyl rubber seals should be used for cultivating the most fastidious anaerobes because of the impermeability of this type of rubber to oxygen (Hungate, 1969). The seals supplied by Wheaton Scientific have a slot moulded in the inner face which enables the gassing jet to be withdrawn easily from the bottle while inserting the seal, whereas those supplied by Pierce and Warriner have no such slot, are made of harder rubber and fit more tightly. Both types of seal have proved satisfactory. The aluminium caps which hold the seals firmly in place are compressed on to the necks of the bottles using a hand crimper (Pierce and Warriner, UK, Ltd).

Preparation of Media and Inoculation

The Hungate technique (Hungate, 1950, 1969; Bryant, 1972) and the serum bottle modification (Miller and Wolin, 1974) have been described in detail elsewhere. Brief outlines of media preparation and inoculation are given here in order that the two methods may be compared.

Hungate technique

A solution of medium ingredients (excluding reducing agents and sodium carbonate) is prepared in water and, after any necessary adjustment of pH, is boiled for one min in a round-bottomed flask under a stream of O_2-free CO_2. The flask is removed from the heat and as the gassing jet is withdrawn a rubber stopper is inserted into the neck of the flask and wired in place. Separate solutions of the reducing agents and sodium carbonate are treated similarly. The reducing agents are usually either cysteine-HCl or a mixture of cysteine-HCl and sodium sulphide. The latter is gassed with O_2-free nitrogen to minimize the evolution of hydrogen sulphide. All three solutions are sterilized by autoclaving and when the flasks have cooled (to around 50° in the case of agar media) the appropriate amounts of reducing agent and carbonate solutions are added to the bulk medium flask. This is done aseptically and with continuous gassing of all solutions with O_2-free gas. The complete medium is dispensed by pipette to sterile test tubes closed with sterile butyl rubber stoppers. Maintaining complete anaerobiosis during the aseptic distribution of media is by far the most difficult part of the technique and the handling of stoppers, pipette and gassing jets requires considerable dexterity. This applies equally to inoculation of the medium when the stoppers have to be temporarily removed.

Serum bottle modification

As with the Hungate technique a solution of the main medium ingredients and separate solutions of the reducing agent(s) and sodium carbonate are boiled under a continuous steam of O_2-free gas (Fig. 1). The medium solution and carbonate solution are then chilled in ice water to approximately room temperature (50° with agar media) and thoroughly equilibrated with O_2-free CO_2. Appropriate volumes of the reducing agent and carbonate solutions are added to the medium which is

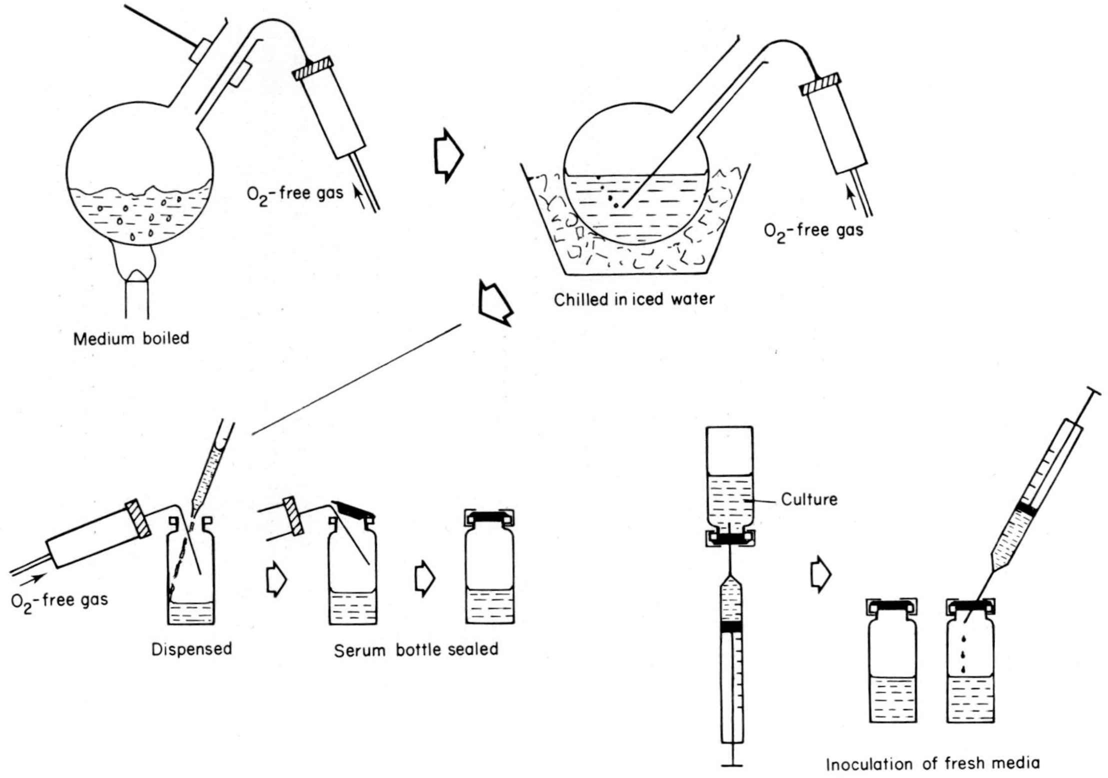

FIG. 1. Flow diagram for preparation and inoculation of media using the serum bottle modification of the Hungate technique. The serum bottle was sealed and autoclaved.

mixed and, while still being gassed, dispensed either by mouth pipette or a continuous pipetting syringe to the serum bottles. During filling, the serum bottles can be held in a rack and a manifold of jets used to gas several bottles simultaneously. Gas is bubbled through the medium in the serum bottles for a short time (10–15 s for a 15 ml bottle filled with 10 ml medium) and a butyl rubber seal pressed into the neck of the bottle as the gassing jet is removed. Aluminium caps are immediately crimped on to the bottles and the bottles containing partially reduced media are then sterilized by autoclaving. The process of autoclaving fully reduces the medium as indicated by the colour of the redox dye, either resazurin or phenosafranine, incorporated in the medium.

Serum bottles are inoculated using disposable graduated 1 ml syringes fitted with 3/8 in, 26 gauge needles (Gillette, Surgical, Ltd, Great West Road, Isleworth, Middlesex). The syringes are aseptically flushed with CO_2 before the cell suspension for inoculation is taken up. The usual inoculum is 0·1 ml to 10 ml broth medium and agar slopes are inoculated by using longer syringe needles and stabbing into the agar. Several bottles may be rapidly inoculated in quick succession.

Advantages of the Serum Bottle Modification for the Culture of Strict Anaerobes

Compared with the Hungate technique the serum bottle modification has several major advantages. For example the necessity to distribute highly reduced media aseptically with its attendant risks of contamination and oxidation is avoided. Consequently, relatively inexperienced workers can readily prepare media sufficiently reduced to allow cultivation of the most fastidious anaerobes. In addition the time taken to prepare and distribute media is considerably reduced. Whereas, by the so-called "open tube" Hungate technique approximately 2½ h is required to prepare and dispense 1 litre of medium in 10 ml portions, the same amount of medium would be prepared and dispensed using the serum bottle method in a little over one-half that time. It should be noted however that similar time savings can be achieved using a "closed tube" Hungate technique of media preparation (Holdeman and Moore, 1972). The serum bottle method also lends itself to automation since the medium is only partially reduced when dispensed. Therefore, provided all the air within the serum bottles is displaced by O_2-free gas before sealing and crimping, no problems of incomplete medium reduction during autoclaving should occur.

Complete anaerobiosis of medium and inoculum is maintained during inoculation and several bottles may be rapidly inoculated in turn.

Routine supplementation of a culture with additional substrate, such as hydrogen for the methanogenic bacteria is easily achieved by injection through the seal.

Possibly the greatest advantage of the method is the ease with which gas production can be measured and samples of headspace gas or culture taken during incubation. The volume of gas produced by a culture is measured readily by means of a syringe (Fig. 2). The culture bottle is shaken to displace dissolved gas from the medium and allowed to equilibrate to room temperature. A graduated syringe, previously flushed with O_2-free gas and fitted with a 3/8 in, 25-gauge needle is inserted through the seal and the pressure of gas within the bottle is allowed to displace the syringe plunger by an amount equivalent to the volume of gas produced. Gas production has been monitored continuously during in-

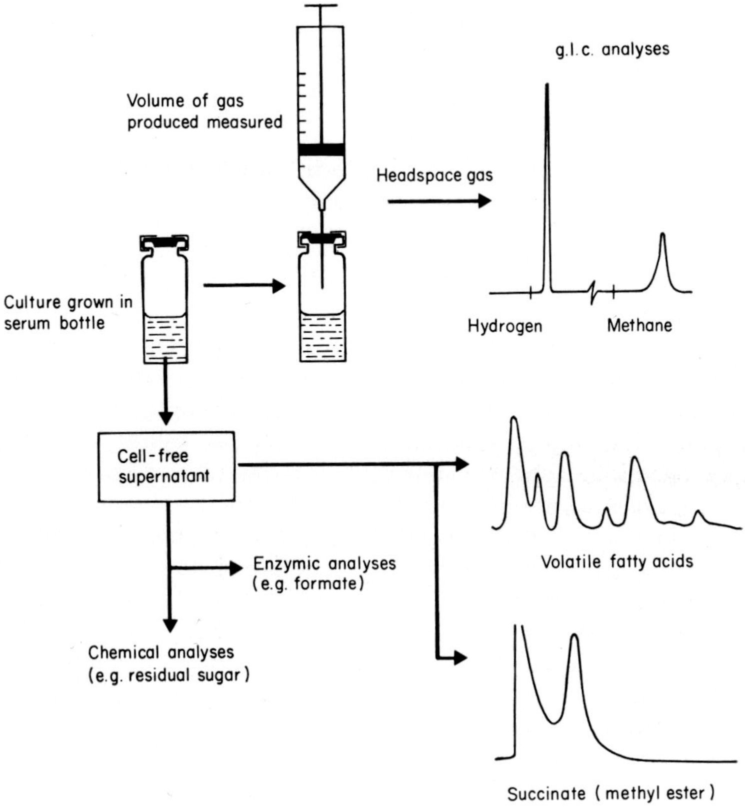

FIG. 2. Flow diagram for analysis of fermentation products of a culture grown in a serum bottle.

cubation by leaving a syringe permanently inserted in the seal with both the syringe and serum bottle clamped firmly in position. So far this technique has proved practicable for relatively vigorous gas-producing organisms, e.g. *Megasphaera elsdenii* grown on lactic acid, but has not been successful when used for measuring hydrogen uptake by methanogenic bacteria possibly because of the permeability of the plastic syringe walls to air. Gas-tight glass syringes might be preferable for this application. The amount of, say, hydrogen or methane, produced by a culture is obtained from analysis of 0·2 to 1·0 ml of culture headspace gas, the gas being drawn into and expelled from the syringe several times before taking the sample. Other fermentation products are determined from the cell-free supernatant obtained after centrifugation of 1 ml samples of culture fluid. Again, withdrawing the sample by syringe can be achieved quickly, aseptically and without danger of oxidation.

Application of the Serum Bottle Technique to Studying Interactions between Anaerobes

The ease with which measurements of gaseous fermentation products can be made makes the serum bottle technique especially suitable for studying interactions involving the production and utilization of gases. This technique has been applied to the study of interactions between the important rumen cellulolytic bacterium, *Ruminococcus flavefaciens* and the methanogenic rumen organism *Methanobacterium ruminantium* when grown in mixed culture (Latham and Wolin, 1977). Details of the culture media used to grow these species and analysis of their fermentation products are given in Appendix 1 on p. 122.

In these experiments the ability to follow methane formation by periodic headspace sampling throughout the incubation period was particularly valuable as a means of detecting the growth of the methanobacterium in the presence of the ruminococcus.

In pure culture *R. flavefaciens* reoxidizes most of the NADH arising during glycolysis by the formation of succinate from oxaloacetate and fumarate, and only a small amount is reoxidized by the formation of molecular hydrogen, H_2. *Methanobacterium ruminantium* derives its energy for growth solely by the production of methane (CH_4) from H_2 and CO_2 or formate. In view of the almost complete absence of H_2, succinate, ethanol and lactate in the rumen, Hungate (1966) suggested that for the mixed population of rumen bacteria, methanogenesis might provide an alternative and energetically preferable means of disposing of reducing power. To test this hypothesis, the ruminococcus was grown in serum bottles as a pure culture and as a mixed culture with the methane bacterium under an atmosphere of 100% CO_2 on a cellulose broth medium. The cultures were incubated on a shaker at 37° until the digestion of cellulose was complete. In pure culture the ruminococcus produced succinate as its major reduced product and only small amounts of H_2 (Table 1). In mixed culture, however, CH_4 was formed by the methanobacterium but in amounts considerably greater than could be accounted for by the small amount of H_2 previously formed by the

TABLE 1. Fermentation products of *Ruminococcus flavefaciens* grown on cellulose* in the presence or absence of *Methanobacterium ruminantium*

	ruminantium *R. flavefaciens* alone (mol 100 mol^{-1} C_6)	*R. flavefaciens* + *M. ruminantium* (mol 100 mol^{-1} C_6)
Acetate	107	189
Formate	62	1
Succinate	93	11
H_2	37	0
CH_4	0	83
CO_2	− 48	94
Carbon recovery (%)	85	85
O/R index	0·87	1·07

* Cellulose equivalent to a mean value of 7·7 mM hexose.

ruminococcus in pure culture. In addition the ruminococcus formed more acetate and less succinate in mixed culture. Thermodynamic considerations strongly suggest that the methanogenic organism, by continually removing the H_2 produced by the ruminococcus and maintaining a low partial pressure of hydrogen in the headspace gas, stimulates further reoxidation of reduced NAD by the formation of molecular H_2. Since less fumarate, and as a consequence less pyruvate, is required as an electron acceptor, the pyruvate so spared is converted to acetate, probably giving rise to an increased yield of ATP to the ruminococcus. Interactions of a similar nature between methanogenic bacteria and the rumen bacterium *Selenomonas ruminantium* (Scheifinger *et al.*, 1975) were also studied using the serum bottle technique and it was found that hydrogen production by the selenomonas was stimulated almost 100-fold by growth with a methanogenic organism. These two H_2 transfer reactions and other interactions between rumen bacteria were recently discussed by Wolin (1975).

As a result of the successful application of serum bottle culture to the study of H_2 transfer reactions, the technique is now being applied to studies of interactions between cellulolytic and non-cellulolytic bacteria involving competition for supply of soluble nutrients, with the advantage that even with small culture volumes the amounts of a variety of soluble energy yielding substrates and growth factors can readily be monitored during growth without disturbing the anaerobic environment.

Conclusions

Many species of bacteria which have previously been isolated and cultured using the Hungate technique grow well using the serum bottle method. These species include the most fastidious of the anaerobes, the methanogenic bacteria, *M. ruminantium*, the MoH symbiont of the culture known as "*Methanobacillus omelianskii*" and *Methanosarcina barkeri*, the cellulolytic rumen bacteria *Ruminococcus albus*, *R. flavefaciens* and *Bacteroides succinogenes* and the non-cellulolytic rumen bacteria *Bacteroides ruminicola*, *Butyrivibrio fibrisolvens*, *Selenomonas ruminantium* *Megasphaera elsdenii*. Thus, there is no reason to suppose that other anaerobes may not be cultivated by this method.

In using this method the ability to remove samples of culture headspace gas or culture liquor rapidly, aseptically and anaerobically has been found to be particularly advantageous in the study of microbial interactions and the ease with which highly anaerobic media can be prepared by the serum bottle method would seem to make its use attractive to those laboratories with little or no experience of the Hungate technique.

References

Bryant, M. P. (1972). Commentary on the Hungate technique for culture of anaerobic bacteria. *Am. J. Clin. Nutr.*, **25**, 1324.

Bryant, M. P. & Robinson, I. M. (1961). An improved non-selective culture medium for ruminal bacteria and its use in determining diurnal variation in numbers of bacteria in the rumen. *J. Dairy Sci.*, **44**, 1446.

Hobson, P. N. & Shaw, B. G. (1974). The bacterial population of piggery-waste anaerobic digesters. *Water Res.*, **8**, 507.

Holdeman, L. V. & Moore, W. E. C. (1972). *Anaerobe laboratory manual.* Blackburg, Virginia, USA: Virginia Polytechnic Institute and State University.

Hungate, R. E. (1950). The anaerobic mesophilic cellulolytic bacteria. *Bacteriol. Rev.*, **14**, 1.

Hungate, R. E. (1966). *The rumen and its microbes.* New York and London: Academic Press.

Hungate, R. E. (1969). A roll tube method for cultivation of strict anaerobes. In *Methods of microbiology*, Vol. 3B (Norris, J. R. & Ribbons, D., eds). London and New York: Academic Press, p. 117.

Kistner, A. (1960). An improved method for viable counts of bacteria of the ovine rumen which ferment carbohydrate. *J. gen. Microbiol.*, **23**, 565.

LATHAM, M. J. & SHARPE, M. E. (1971). In *Isolation of anaerobes*. Soc. appl. Bact. Tech. Ser. No. 5. London and New York: Academic Press, p. 133.
LATHAM, M. J. & WOLIN, M. J. (1977). Fermentation of cellulose by *Ruminococcus flavefaciens* in the presence and absence of *Methanobacterium ruminantium*. *Appl. environ. Microbiol.*, **34**, 297.
MAH, R. A. & SUSSMAN, C. (1968). Microbiology of anaerobic sludge fermentation. 1. Enumeration of the non-methanogenic bacteria. *Appl. Microbiol.*, **16**, 358.
MILLER, T. L. & WOLIN, M. J. (1974). A serum bottle modification of the Hungate technique for cultivating obligate anaerobes. *Appl. Microbiol.*, **27**, 985.
SCHEIFINGER, C. C., LINEHAN, B. & WOLIN, M. J. (1975). H_2 production by *Selenomonas ruminantium* in the absence and presence of methanogenic bacteria. *Appl. Microbiol.*, **29**, 480.
WOLIN, M. J. (1975). In *Digestion and metabolism in the ruminants*. New England: Univ. of New England Publishing Unit.

Appendix 1

Media

Cellulose medium for growth of *Ruminococcus flavefaciens* (100 ml^{-1} medium)

cellulose	0·2 g
trypticase	0·5 g
yeast extract	0·1 g
mineral Solution I	7·5 ml
mineral Solution II	7·5 ml
resazurin solution	0·1 ml
branched VFA mixture	0·1 ml.

Ingredients, except cellulose and the sodium carbonate and cysteine HCl-sodium sulphide solutions, were dissolved in water, the pH adjusted to 6·8 with 1 M NaOH and the volume made up with water to 93 ml. The medium was boiled and chilled, whilst being gassed continuously with CO_2.

Sodium carbonate solution	5·0 ml
cysteine HCl-sodium sulphide solution	2·0 ml.

After addition of the sodium carbonate and cysteine HCl-sodium sulphide solutions the medium was mixed and distributed to gassed serum bottles containing cellulose (Whatman No. 1 filter paper either as a ball milled suspension in water or hammer-milled powder). The bottles were sealed, capped and crimped, and autoclaved at 108 kPa for 15 min.

Medium without rumen fluid for growth of *Methanobacterium ruminantium* strain PS (100 ml^{-1} medium)

trypticase	0·4 g
yeast extract	0·2 g
sodium formate	0·2 g
sodium acetate	0·2 g
mineral Solution I	7·5 ml
mineral Solution II	7·5 ml
ferrous sulphate solution	0·1 ml
resazurin solution	0·1 ml
sodium carbonate solution	2·5 ml
cysteine HCl-sodium sulphide solution	3·0 ml.

The medium was prepared as for the cellulose medium but distributed to serum bottles under 20% H_2/80% CO_2.

Media solutions

Mineral Solution I

K_2HPO_4	0·6% (w/v) in distilled H_2O.

Mineral Solution II

NaCl	1·2% (w/v)
$(NH_4)_2SO_4$	1·2% (w/v)
KH_2PO_4	0·6% (w/v)
$CaCl_2$ anhydrous	0·12% (w/v)
$MgSO_4{\cdot}7H_2O$	0·25% (w/v)

all dissolved in distilled H_2O.

Resazurin solution

0·1% (w/v) in distilled H_2O.

Branched volatile fatty acid (VFA) mixture

iso-butyric acid	1 volume
DL-2-methyl butyric acid	1 volume
iso-valeric acid	1 volume.

Sodium carbonate solution

sodium carbonate	8·0% (w/v)

boiled, chilled and equilibrated with CO_2.

Cysteine HCl-sodium sulphide solution

cysteine-HCl	1·25% (w/v)
sodium sulphide 9 H_2O	1·25% (w/v)

Cysteine dissolved in water, pH adjusted to 10·0 with 10 M NaOH, then sodium sulphide added. Boiled and gassed with nitrogen.

Ferrous sulphate solution

$FeSO_4 \cdot 7H_2O$ 1·0% (w/v) in 1·0% HCl (v/v).

Analytical methods

Hydrogen and methane

Hydrogen and methane were determined from 0·2 to 1·0 ml samples of culture headspace gas by gas chromatography (Pye 104 chromatograph, katherometer detector; Pye Unicam Ltd, Cambridge) using a 5 ft × $\frac{1}{4}$ in glass column packed with silica gel 100 mesh and using nitrogen (50 ml min^{-1}) as the carrier gas.

Volatile fatty acids and succinic acid

Volatile fatty acids were determined by gas chromatography using a Perkin Elmer 900 fitted with a flame ionization detector (Perkin Elmer Ltd, Beaconsfield, Buckinghamshire). Culture supernatant, 0·6 M orthophosphoric acid and a 0·25% (v/v) solution of *n*-pentanol as internal standard were mixed in the ratio 10:5:1 by volume. Components in 2 μl injections were separated on a 6 ft × $\frac{1}{8}$ in stainless steel column packed with Chromosorb 80–100, polyethylene glycol 400 and orthophosphoric acid (18:2:0·1) at 115° with nitrogen (35 ml min^{-1}) as the carrier gas. Succinate, as its methyl ester, was also determined by gas chromatography using the same instrument on a column packed with Diatomite C coated with 20% (w/w) polyethylene glycol adipate at 150° with nitrogen (35 ml ml^{-1}) as the carrier gas.

Formic acid

Formic acid was determined enzymically by the method of Rabinowitz and Pricer (1965).

References

RABINOWITZ, J. C. & PRICER, W. E. (1965). Formate. In *Methods of enzymatic analysis*. New York and London: Academic Press.

Anaerobic Bacteria in Mixed Cultures; Ecology of the Rumen and Sewage Digesters

P. N. Hobson and R. Summers

The Rowett Research Institute, Bucksburn, Aberdeen, Scotland

Introduction

This paper is not intended to be a comprehensive review of rumen and digester microbiology; rather it tries to set out, with some examples of the various techniques, an approach to the investigation of the ecology of particular microbial habitats which have been used in our laboratories and which is applicable to other microbial systems. For more detailed considerations of the subject including references to work in other laboratories, see the review papers by Hobson (1969a, 1971, 1972, 1973, 1976); Hobson *et al.* (1974) and for the rumen alone, Hobson and Howard (1969) and Hungate (1966).

The Habitats

Both the rumen and high-rate sewage digesters are essentially the same when considered as microbial habitats, in that they are stirred, semi-continuous, highly-anaerobic cultures at mesophilic temperatures, containing a very mixed population of bacteria (and protozoa in the rumen), in large numbers fed with mainly solid material of high carbohydrate, protein and lipid content. The two systems are shown in Figs 1a, b. The rumen is stirred by muscular contractions; it is heated by metabolic heat from the animal and from the microbes, and the overflow is controlled by pressure of the digesta and by the size of the digesta particles. Some metabolites are absorbed through the rumen wall while others flow out with the overflow and fermentation gases are eructated from the ruminant's mouth. The sewage digester is stirred by mechanical means; it is heated by a pumped system indirectly using the fermentation gas and the overflow is by gravity and pressure from the input or by pumping. Metabolites flow out with this overflow and the fermentation gas is piped from the top of the digester.

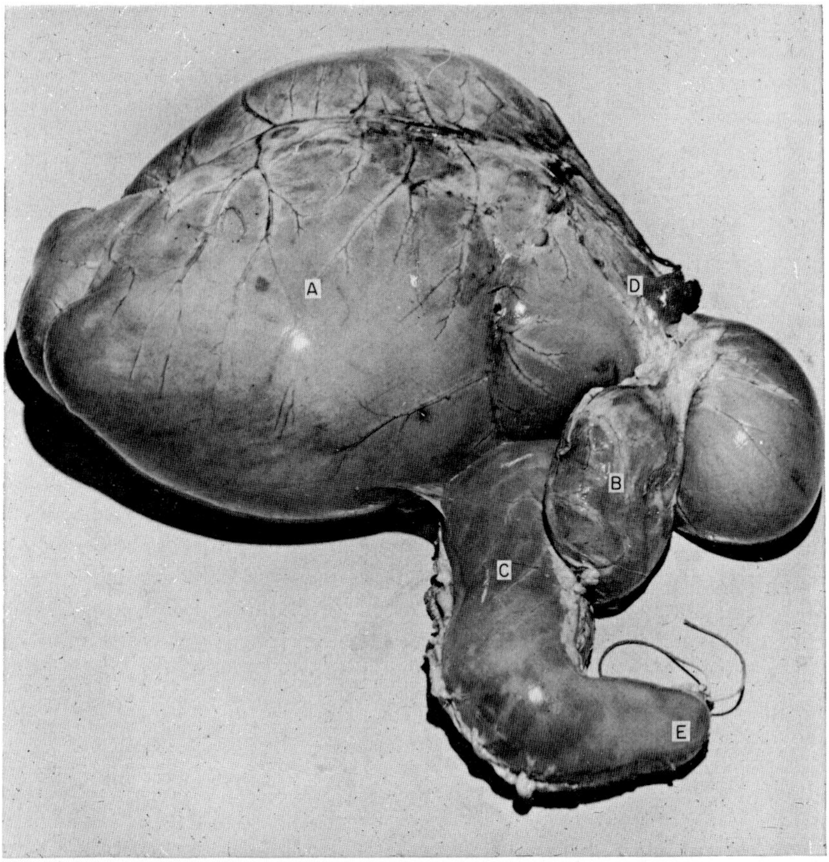

FIG. 1. (a) The stomach system of a deer showing reticulo-rumen compartments A, omasum B, abomasum or true stomach C. Food enters at D and passes from A to B to C and then to the small intestine at E. The capacity of A varies from about 5 litres in a sheep to 150 litres in a cow.

There are, however, two fundamental differences between the two systems. In the rumen there is a more or less continuous flow of saliva which forms the basal medium for the micro-organisms and which, with the exception of some urea, contains no energy or nitrogen sources for the organisms. Added to this flow of basal medium, at varying intervals depending on feeding practice, are the actual substrates for microbial metabolism. Although some feed particles may remain in the rumen longer, the turnover time for the saliva and for the suspended micro-organisms and small particulate, or dissolved, feedstuffs is about one day, or less. The basal liquid and microbial substrates are added to the

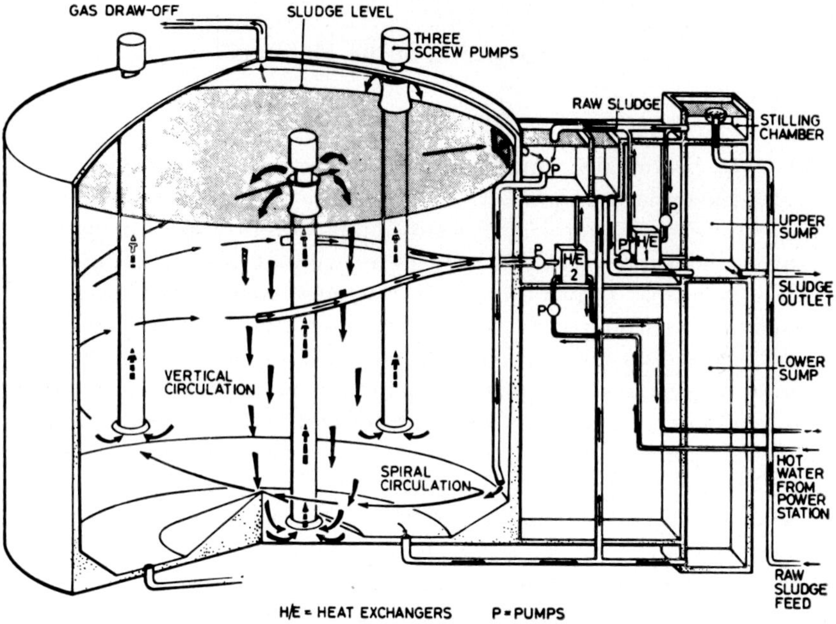

FIG. 1. (b) Diagram of the microbiologically similar sewage digester, showing one system for heating and stirring. The capacity of digesters can be 4·5 million litres or even more. Reproduced by courtesy of *Process biochemistry*.

sewage digester at the same rate, and although some larger particles may not flow out at the same rate as the liquid, depending on the stirring, the overall turnover time of the liquid and suspended matter is a minimum of about 10 days and more often about 30 days. The large difference in nominal dilution rate is probably a cause of some of the differences in metabolic activities in the two systems.

There is also a difference in the objectives of the two systems and it is towards maximizing the efficiency of the systems that microbiological research is ultimately directed, as both systems are commercially useful. In the rumen system a maximum conversion of feed into fermentation acids and microbial cells is desired, as these are what form the host animal's food. Minimum production of gas is desired as this is of no use to the animal. In the digester minimum residual fermentation acids and microbial cells are wanted, as these form bulk and pollutants in the overflow and the whole point of the system is to reduce these factors in the incoming sewage. Maximum production of gas is required as the gas is non-polluting and commercially valuable.

Investigation of the Habitats

Chemical analysis

Although, historically, microbes were observed in the rumen before chemical analysis of the system was possible, in the case of both the rumen and digester biochemical analyses of input, output and contents of the systems and experiments *in vitro* have, without any reference to the micro-organisms, shown the main reactions occurring. However, although input-output analysis may by observation and trial and error enable suggestions for improving the efficiency of the system to be made, no basic suggestions, or explanations, can be made without consideration of the micro-organisms and their metabolism. Without adequate microbiological control, errors may occur in interpretation of biochemical tests. For a complete analysis of the system both chemical and microbiological, observations and experiments must be made and correlated. A comparison of the main biochemical pathways in the rumen and digester is given by Hobson *et al.* (1974).

Microscopic observation

This shows that the microbial system is complex and that a diversity of morphological forms of micro-organisms are present. It may show that some morphological forms are apparently connected with the breakdown of particular substrates which can be microscopically identified; e.g. starch granules, leaf or paper fibres. It can show any large variation in proportions of different morphological forms with variation in amount or type of input to the system, pH of the system, accumulation of some metabolite or gross derangement of the system. It can tell nothing about the reactions of the organisms. However, as the first method of analysis of the habitat, microscopic observation of the whole system remains a necessary adjunct to further microbiological work and with advances in work *in vitro* more detail may be added to what was previously purely observation of morphological forms. Pleomorphism and the difficulties in observing small changes in a mixed microbial population can, however, limit the usefulness of direct microscopy.

Batch culture and counting

In the analysis of a microbial habitat one needs to know how many and what types of micro-organisms are present, so culture of the organisms is an indispensable step. Culture of bacteria depends on suitable media and it need hardly be said that only the bacteria which grow in a particular medium will be isolated or counted. This can give wrong impressions of the predominant bacteria and it was not until specialized media and methods had been developed that a true assessment of the rumen and digester anaerobic flora was possible (for further information on media and methods see Hobson, 1969b; Hungate, 1966, 1969; Shapton and Board, 1971).

As a first step towards assessing a flora, media designed to grow all the bacteria in the system are often used. The bacteria growing in these media can then be counted, and isolated for secondary culture. Secondary culture enables the bacteria of the habitat to be classified according to their reactions and may identify those species with particular morphological forms observed in the habitat. It can give relative concentrations to species and show how the concentration of a species varies with time or with conditions developing in, or imposed on, the system and it provides additional bacteria for further experiments.

A count of bacteria growing in a general medium, designed from chemical analysis of the habitat to grow all bacteria, may indeed, give a reasonable approximation of the numbers of viable bacteria in the system and such media are used to report a complete or total viable count for the system. However, for some analytical purposes the number of bacteria with particular biochemical properties (which may be single species or groups of species) is of more importance than the whole number. For such an analysis recourse may be had to selective media or media which give a visible reaction with bacteria having the particular biochemical property. The use of such media gives a direct count and is less time-consuming than the use of general media and subsequent secondary culture and biochemical testing of isolates as a method of assessing numbers of bacteria with particular properties.

A further method of enumerating the micro-organisms of the system is by total counts, where the organisms are counted microscopically or by electronic methods (see Hobson and Mann, 1970). This method can be used to count all organisms or particular morphological groups. It has the advantage that it does not depend on the growth of organisms removed from the habitat, but the disadvantage that it cannot indicate the metabolic status of the organisms.

Only bacteria have been mentioned so far, as we have never observed protozoa in sewage digesters. Although the rumen protozoa can be cultured (culture of the ciliate protozoa is described in the paper on p. 143 of this volume), culture is only possible non-axenically and with difficulty in undefined media and it is not possible to use culture of protozoa as a method of counting as for bacteria, or to use cultures in the many ways in which bacterial cultures are used.

The rumen protozoa may only be assessed by total counts, that is counts of all organisms in fixed preparations. With the rumen protozoa viable organisms are probably a large proportion of the total. With rumen or digester bacteria the viable counts are nearly always only a small fraction (perhaps 10% or less) of the total counts. While there are reasons for predicting, because of the nature of bacterial growth, that many of the bacteria will be dead, the situation is complicated by the fact that the terms viable and non-viable are to some extent artefacts defined by whether a bacterium will produce a colony when transferred from its natural habitat to an artificial medium. Some bacteria which grow in the habitat may not grow in the culture when first transferred. In addition it is known that some species of bacteria will not grow in the media in which the majority of bacteria from the habitats will grow and which are designed for complete viable counts. Such bacteria are usually only a small percentage of the bacterial population. In addition, however, it is possible that so-far uncultured and unidentified bacteria exist in habitats such as the rumen where the known species and strains of bacteria number several hundreds. A further complication of both total and viable counts of organisms in such habitats as the rumen is that the organisms are attached to or entrapped in the solids in the habitat contents. No effective way of removing all such organisms exists. However, if samples are treated by a standard procedure the bacteria counted should be a constant proportion of the total and so counts can be used to show relative changes in the populations of the habitat. Thus, data can be obtained showing diurnal changes in rumen flora with time after feeding, changes with season (i.e. with food available), longer-term changes with change in feed of growing animals and so on. By using the selective media previously mentioned relative changes in functional groups of bacteria may be described. An example is the development of the methane-producing flora in a digester starting from piggery waste (Hobson and Shaw, 1974). Examples of the use of viable counts in monitoring diurnal and longer-term changes in the whole, and particular groups or species of, rumen bacteria are given by Kurihara *et al.* (1968) and Eadie *et al.* (1959). An example of the use of total counts to investigate interactions between morphologically distinct groups of rumen organisms is also given by Kurihara *et al.* (1968), who present graphs showing the growth, over some weeks, of a protozoal population from a small inoculum in a previously unfaunated rumen and the decrease in numbers of bacteria due to predation by the protozoa, until an equilibrium state was reached.

An additional factor which must be taken into account in the examination of a continuous or semicontinuous natural culture is the dilution rate of the system. It may be possible to measure this easily, but in the rumen where salivary secretions of the animal form a major part of the flow indirect measurement by determining the rate of change of concentration of markers added to the rumen contents must be used.

Culture can be used as a method of measuring changes in the microbial activity of the habitat, but to obtain fullest information on the meaning of these changes the reactions of the bacteria must be known. Batch cultures can show the substrates used and fermentation products and enzymes produced by particular bacteria. These can be compared with those detected in the natural habitat or produced *in vitro* by cultures or washed suspensions of the mixed flora of the habitat. Cultures can also show nitrogen sources, growth factors or trace elements used by the bacteria. This may need the use of isotopically labelled compounds: an example of the use of ^{15}N labelling is given by Hobson *et al.* (1968).

Once the components of the microbial population and their reactions have become known pure batch cultures may be used, for example, to test the actions of possible inhibitors on bacteria selected (on the basis of their numbers and reactions) for their presumed importance in the system. An example is given by el Akkad and Hobson (1966) of the actions of antibiotics on major members of the rumen flora. Antibiotics were at that time added to feedstuffs to improve animal performance, and might still get into the rumen or farm waste digesters when used prophylactically for animals. Heavy metals, ammonia and volatile fatty acids may inhibit anaerobic digestion. Results of experiments (Shaw, 1971) testing inhibition of a major methanogenic bacterium by different concentrations of copper, ammonia and lower volatile fatty acids suggested that methanogenesis was unlikely to be inhibited by the concentrations of these substances found in piggery waste. Tests on actual digesters have confirmed these conclusions. The work of Henderson (1973) on the effects of long-chain fatty acids on the growth of species of rumen bacteria selected for their biochemical activities helped to explain and predict the actions of some substances in animal feedstuffs which alter, or could alter, the rumen fermentation. Fats are constituents of normal ruminant diets and extra fat may be added to some types of diet, and the glycerides are rapidly hydrolysed in the rumen to free fatty acids. Other pure culture studies on hydrogen uptake by rumen bacteria (Henderson, 1974) have helped to elucidate one theory of rumen function examined from the opposite aspect in mixed continuous culture.

Ancillary methods

Other methods may be of use in particular problems associated with the analysis of a mixed culture. Two examples are the use of serological methods with fluorescent antibodies, and micromanipulation. However, the first method requires cultural isolation of particular bacteria so that antisera may be prepared. The second also requires media in which the single, isolated organisms may be presumed to be able to grow. The use of serological techniques may be limited to some extent, as in rumen analysis, by the fact that many serological types of biochemically distinct species of bacteria occur, and the balance of these serological types varies with time. This latter fact may of itself indicate continued inoculation of the rumen from outside sources (Hobson *et al.*, 1958). The fluorescent antibody technique may be used for rapid screening of rumen samples for particular types of bacteria, screening of intestinal contents of different animals for rumen-type bacteria, comparison of cultured bacteria with morphologically differing bacteria in the rumen, "tracking" a particular type of bacterium introduced artificially into the rumen and noting association of particular bacteria with particular feed debris in the rumen (Hobson and Mann, 1957, 1961; Hobson *et al.*, 1955, 1958, 1962).

The use of micromanipulation in attempts to solve the problems of some apparently "unculturable" cocci and the discrepancy in size between selenomonads *in vivo* and in culture was illustrated by Purdom (1963).

Continuous Pure Cultures

During the growth of batch cultures the bacteria are subject to varying conditions of growth rate, substrate concentration, pH, end product concentration and bacterial concentration. These may all affect bacterial metabolism and the final analysis of fermentation products may be different from the production of the bacteria in their natural habitat. Continuous culture allows the growth of bacteria under more defined conditions and, as the rumen and digester are forms of continuous culture, approximates more closely to the natural conditions.

Several examples of the use of continuous pure culture in rumen microbiology have been published (Henderson *et al.*, 1969; Hobson, 1965; Hobson and Smith, 1963; Hobson and Summers, 1966, 1967, 1972). The bacteria used in the cultures were selected because other methods of culture and counting had shown them to be important members of most rumen floras, and because they also had many properties representative of those of the rumen flora. The importance of growth rate in determining growth yield, fermentation products and enzymic activities is described in the papers cited above and the effects of the maintenance requirement of bacteria in C- and N-limited cultures is discussed. Although the rumen or anaerobic digester is a continuous culture the system does not behave as a simple chemostat. The overall growth rate of the bacteria must conform to the overall dilution rate of the culture but, because of the method of feeding nutrients to the system, the growth rate at any particular time is governed by the availability of nutrients. The addition of one feed batch is followed by a burst of activity by the bacteria and as nutrients are used up this is followed by a period of slow growth, or a stationary phase and even cell lysis, until the next feed input. Thus, the systems tend towards a series of batch cultures imposed on a continuous culture. Since the feeds are complex and contain much polymeric material, substrates for the bacteria become available at different rates and at different times after the feed, and the overall "batch culture" described by total, or complete-viable, counts of bacteria in the systems can be shown by analysis of particular groups or species of bacteria to be made up of a series of batch cultures of bacteria with different nutrient requirements. Diauxic growth may also occur. In the rumen the protozoa also have a sequence of division in relation to feeding time of the animal different from that of the bacteria. However, the different microbial growth and the metabolic pathways of the whole system must be in equilibrium. For instance in either the rumen or digester the feeding of easily fermented substances in large amount can give rise to rapid microbial activity which overwhelms the normal metabolic pathways of the mixed culture and leads to excess acidity which can cause death of the animal or "souring" and failure of a digester.

Using the data obtained from continuous pure cultures the sequence of events following overfeeding a ruminant can be predicted and explained, as can other observations on rumen function (Hobson, 1972). The data can also be used to calculate the quantitative behaviour of a rumen system given relatively simple feedstuffs and quite good agreement with results of metabolic experiments *in vivo* is obtained. The objective of a rumen system is to obtain maximum conversion of feed to microbial cells and volatile acid fermentation products on which the ruminant lives, with minimum production of CH_4 which is a waste product to the animal. Consideration of results *in vitro* shows that the feeding of ruminants is rarely, if ever, done in such a manner as to bring about this maximum efficiency. On the other hand the objective of a digester system is to bring about maximum efficiency in the opposite direction, i.e. minimum production of bacteria (residual sludge), minimum production of acidic fermentation products which increase BOD and COD of the output and maximum conversion of feed to CH_4 which is valuable as an energy source. Ways of increasing these efficiencies can be suggested from microbiological data, although these have to be within the bounds of commercial restraints.

Batch and Continuous Mixed Cultures

So far metabolic experiments on pure cultures have been described but the microbial systems being investigated are mixed cultures which are too complex for complete ecological analysis. Pure cultures can suggest how interactions of micro-organisms may occur in the complete systems but many aspects of interactions can only be determined by using mixed cultures. It is possible to make detailed analyses of cultures containing two, or possibly three, components and mixed two-component cultures have been used in investigations of rumens and digesters, especially the former. Some aspects of cultures showing interactions of rumen protozoa with bacteria are discussed elsewhere (in the paper on p. 143 of this volume).

Batch cultures were used (Hobson and Stewart, 1970) to demonstrate competition between *Selenomonas ruminantium*, which occurs regularly in the rumen flora, and a lactobacillus species which can dominate the flora when large amounts of easily fermented carbohydrate are fed and the rumen becomes acid. The change in pH after feeding is about 6·5 to 5·5 or less on such diets compared with about 6·9 to 6·5 on roughage diets. In culture at pH 6·7 minimum doubling times were 90 min for the selenomonad and 150 min for the lactobacillus, but as the pH fell the growth rate of the selenomonad decreased most rapidly and in mixed cultures of different pH the lactobacillus dominated the flora at low pH. It was concluded that the pH changes consequent upon a rapid fermentation in the rumen could explain the occasional dominance of lactobacilli.

A large proportion of the rumen bacteria depend completely or in part on ammonia as a nitrogen source. Although much of this ammonia is usually supplied from degradation of feed proteins, urea is always present in ruminant saliva and this urea, through hydrolysis by bacterial urease, contributes to the rumen ammonia pool. The ureolytic bacteria seem to form only a small proportion of the total bacteria and this still applies even when urea forms the major part of the nitrogen in the feed, as when cattle were fed on urea plus molasses with a little roughage (Elias, 1971). With such a diet, most of the cell mass of the rumen bacteria is derived ultimately from ammonia formed by the ureolytic bacteria but even though the principal ureolytic bacteria could use ammonia as nitrogen source they were still only present in small numbers. In unpublished experiments (Elias, Hobson and Summers) pure and mixed continuous cultures of a typical ammonia-utilizing bacterium (*Bifidobacterium bifidus*) and a major ureolytic bacterium (*Peptostreptococcus* sp.) from the urea-molasses fed animals, whose rumen flora was different from that of animals on normal diets, were run under N-limited conditions with sucrose in excess. These showed that both bacteria would grow in an ammonia medium but that, despite the evidence from studies of growth rates against ammonia concentrations that the *Bifidobacterium* should be dominant (Fig. 2), in mixed culture in ammonia

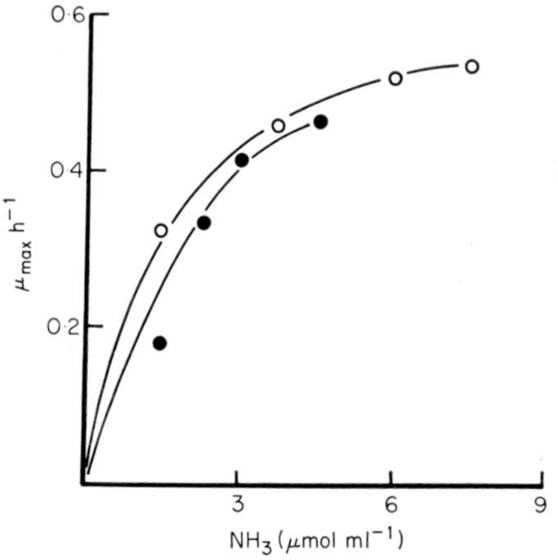

FIG. 2. The dependence of growth rate of rumen strains of *Bifidobacterium bifidus* (○) and *Peptostreptococcus* sp. (●) on ammonia nitrogen concentration in batch culture with sucrose in excess.

medium the bacterium established first remained dominant (cf. results of Meers and Tempest, 1968). There was always some fluctuation in numbers of bacteria in the mixed cultures. An inoculum of *Peptostreptococcus* was added to a culture in which a *Bifidobacterium* population of $5{\cdot}97 \times 10^7$ ml^{-1} had been established at a dilution rate of 0·1 h^{-1}. Over the next 400 h the dilution rate was kept at 0·1 h^{-1} and counts of the bifidobacteria rose and then levelled out, averaging $7{\cdot}82 \times 10^7$ ml^{-1} over the period. However, although counts of peptostreptococci rose to 100×10^3 ml^{-1} at about 90 h after inoculation, they then reduced to an average of 52×10^3 ml^{-1} for the rest of the period. The dilution rate was then changed to 0·065 h^{-1} for 400 h. Over this period the bifidobacteria counts averaged $5{\cdot}89 \times 10^7$ ml^{-1} while the peptostreptococci were still 88×10^3 ml^{-1}.

Bifidobacteria were added to a culture in which peptostreptococci had been established at a dilution rate of $0{\cdot}1\ h^{-1}$ and at numbers of $3{\cdot}74 \times 10^7\ ml^{-1}$. Over the following 500 h at the same dilution rate the bifidobacteria never became more than just visible in Gram-stained films of the culture. This result further demonstrates the impossibility of establishing high populations of a "foreign" bacterium in a stable rumen population *in vivo* as found previously (Hobson and Mann, 1961). However, the *Bifidobacterium* would not grow alone in a urea medium but it would grow when inoculated into a culture of the ureolytic *Peptostreptococcus*, where, although there were some differences in the time scale in different

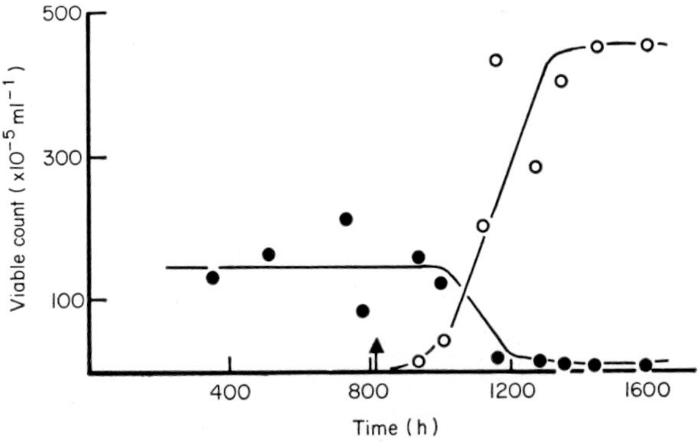

FIG. 3. The growth of the ammonia-utilizing *Bifidobacterium* (○, inoculated ↑) with the ureolytic, ammonia-utilizing *Peptostreptococcus* (●) in a medium containing only urea as N-source with sucrose in excess. From 1600 to 2000 h the relative numbers were similar. Dilution rate $0{\cdot}1\ h^{-1}$. The *Bifidobacterium* would not grow alone in this medium.

experiments, it always eventually became the dominant bacterium. Figure 3 shows the results of one experiment where dominance of the bifid was rapid and without fluctuations in relative numbers. A strain of *Streptococcus faecium* was found to be the principal ureolytic bacterium in roughage-fed sheep (Cook, 1972, 1976), but this bacterium could not use ammonia and needed a complex amino acid mixture for growth. Again, the ureolytic bacterium was only a minor component of the whole flora. In a medium containing amino acids and urea it established stable continuous cultures, but when such a culture was inoculated with the *Bifidobacterium* this became dominant and growth of the *Streptococcus* was reduced in spite of the fact that there was in this case no direct competition for nitrogen source (Fig. 4). Thus, the suggestion from rumen analysis that ureolytic bacteria are a small proportion of the total bacterial mass is confirmed in cultures, but the explanation for this is so far not known.

The work of Iannotti *et al.* (1973) in America provides an example of mixed culture studies applicable to both the rumen and the digester (Hobson *et al.*, 1974). Here the effects, on the production of reduced fermentation acids and hydrogen by carbohydrate-fermenting bacteria, of growth with a hydrogen-utilizing bacterium, such as the methanogenic bacteria of the natural system, were shown. This, with the pure culture work mentioned above, provided experimental verification of an earlier hypothesis on rumen function.

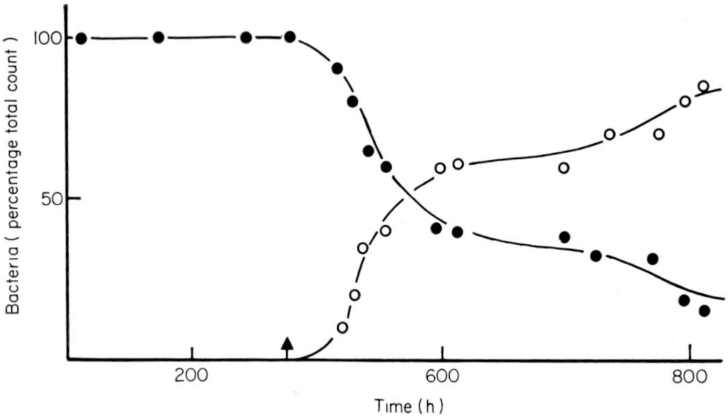

FIG. 4. The growth of the ammonia-utilizing *Bifidobacterium* (○, inoculated ↑) with the amino acid-utilizing, ureolytic *Streptococcus faecium* (●) in a medium containing urea and amino acids as N-sources with glucose as energy source. Dilution rate 0·1 h^{-1}. From 800 to 1000 h the bacteria remained in similar proportion.

Other two-component culture experiments on aspects of rumen bacterial interactions have been done and these, in general, have yielded more information about the rumen micro-organisms than have attempts to culture the whole rumen contents in an artificial rumen. A digester is a completely mechanical system and, within the practical limits of handling the sewage substrates on a small scale, any size of digester should give the same results and contain the same microbial population. On the other hand a rumen is an organ in a living body and it is impossible to reproduce this exactly by a mechanical system. Mixing is different and the overflow of materials from the rumen is through a natural valve which limits the size of particles which can pass out. Acid fermentation products and ammonia are absorbed across the rumen wall by processes which differ from those of the only mechanical substitute, i.e. a dialysis membrane. The basal medium flowing through the rumen is saliva, which can be reproduced in its ionic content for mechanical systems but not in respect of its mucous content. The effects of rumination cannot be reproduced mechanically. Thus, although some artificial rumen systems can simulate fairly well the gross reactions of the rumen over a period of perhaps two or three weeks, no artificial rumen will keep the exact balance of micro-organisms which existed in the rumen from which it was inoculated.

Gnotobiotic Animals

There are thus animal factors involved in rumen microbiology which are impossible to reproduce *in vitro* and which may affect the behaviour of the micro-organisms in their natural habitat. The rumen population of the young animal is acquired, as the animal grows, by inoculation from saliva and other excretions of older animals and from organisms transferred from other animals by air or by food and water. Because of their size and sensitivity to cooling and oxygenation the rumen ciliate protozoa are transferred only over short distances and mainly by animal contact, so by suitable simple isolation it is possible to keep animals free from these protozoa. But it is not possible to prevent the development of a full population of flagellates and bacteria in animals in isolated pens or houses open to all sources of contamination other than direct animal contact. Thus, it is impossible to develop a particular rumen flora in animals kept in normal conditions.

To be able to develop and study known rumen floras in the animal needs gnotobiosis, and we, initially in collaboration with workers at Cambridge, are carrying out such work. During the period of milk feeding a young ruminant behaves as any other mammal and digestion can be carried on in the abomasum and intestines without the aid of micro-organisms. However, the normal solid type of herbivorous diet cannot be digested without the aid of gut micro-organisms and in the normal ruminant the development of the rumen as an organ and the ability of the animal to digest herbage are linked with the increasing intake of solid food which naturally occurs as the young animal grows. The intake of solid food increases from "nibbling" during the early days of life to a quantity sufficient entirely to support the animal at weaning. Physical action of the food (its roughness, etc.) plays some part in rumen development, but the colonization of the solid food by micro-organisms and the production of fermentation acids is also important. It is thus impossible to have a germ-free ruminant, and the ability of an older germ-free animal to live with the undeveloped rumen, on milk or on a non-ruminant diet, seems to be limited.

Although a rumen flora normally contains some hundreds of species and varieties of bacteria and some half dozen species of protozoa, the protozoa are not essential to the rumen function and theoretically the biochemical pathways can be duplicated by a mixture of some nine or ten bacteria. The object of the gnotobiotic lamb experiments has been to see if such a bacterial flora will establish in the rumen with the bacteria in the correct proportions and numbers, and if this flora will carry out the rumen functions in such a way as to lead to growth and health of the lambs. Many of the experimental results are still unpublished, but some are described by Lysons *et al.* (1971, 1976) and Mann and Stewart (1974).

From some points of view the experiments have been successful in that by sequential inoculation of single species or groups of species an initial flora representative of a natural milk-fed lamb can be established and then gradually replaced, in the same way as the natural sequence, by a stable flora representative of the normal adult flora and which carries out the desired reactions. Various aspects of microbial ecology of the rumen and intestines and of rumen functions, have been noted. However, except perhaps in the present experiments where up to the time of writing a predominantly starch-digesting flora seems to be fermenting a starchy-concentrate diet sufficiently well to produce good growth of the lambs, the floras do not seem to have carried out their reactions at a rate sufficient to maintain growth of the lambs beyond a certain age (about 11–12 weeks) and weight. This, especially in the case of a fibre (grass)-digesting flora, has led to a number of questions and experiments about the relevance of tests *in vitro* on purified substrates to the breakdown of the same substrates in a plant material *in vivo*. The successful rearing of lambs with a defined rumen flora and no other organisms will open up many areas of experiments not only on rumen function and the ecology of the commensal rumen and intestinal organisms but also on rumen malfunctions and the interactions of pathogenic and potentially pathogenic bacteria with the commensal flora in the gut.

References

EL AKKAD, I. & HOBSON, P. N. (1966). The effects of antibiotics on the growth of some rumen and intestinal bacteria. *Sond. Zentralblatt. Veterinärmedizin.* **A13**, 700.

COOK, A. R. (1972). Ureolysis in the ovine rumen. *Biochem. J.*, **127**, 66.

COOK, A. R. (1976). Urease activity in the rumen of sheep and the isolation of ureolytic bacteria. *J. gen. Microbiol.*, **92**, 32.

EADIE, J. M., HOBSON, P. N. & MANN, S. O. (1959). A relationship between some bacteria, protozoa and diet in early-weaned calves. *Nature, Lond.*, **183**, 624.

ELIAS, A. (1971). *The rumen bacteria of animals fed on a high-molasses-urea diet.* Ph.D. Thesis, Univ. of Aberdeen.

HENDERSON, C. (1973). The effects of fatty acids on pure cultures of rumen bacteria. *J. agric. Sci., Camb.*, **81**, 107.

HENDERSON, C. (1974). Uptake of extracellular hydrogen by rumen bacteria. *Proc. Soc. Gen. Microbiol.*, **II**, 16.

HENDERSON, C., HOBSON, P. N. & SUMMERS, R. (1969). The production of amylase, protease and lipolytic enzymes by two species of anaerobic rumen bacteria. In *Continuous cultivation of microorganisms* (Màlek, I., ed.). New York and London: Academic Press, p. 189.

HOBSON, P. N. (1965). Continuous culture of some anaerobic and facultatively anaerobic rumen bacteria. *J. gen. Microbiol.*, **38**, 167.

HOBSON, P. N. (1969a). Microbiology of digestion in ruminants and its nutritional significance. In *International encyclopaedia of food and nutrition*, **17**, 59. Oxford: Pergamon Press.

HOBSON, P. N. (1969b). Rumen bacteria. In *Methods in microbiology* (Norris, J. R. & Ribbons, D. W., eds). Vol. 3B. New York and London: Academic Press, p. 133.

HOBSON, P. N. (1971). Rumen microorganisms. *Prog. Ind. Microbiol.*, **9**, 41.

HOBSON, P. N. (1972). Physiological characteristics of rumen microbes and their relation to diet and fermentation patterns. *Proc. Nut. Soc.*, **31**, 135.

HOBSON, P. N. (1973). The bacteriology of anaerobic sewage digestion. *Process Biochem.*, **8**, 19.

HOBSON, P. N. (1976). Trends and innovations in rumen microbiology. In *Microbiology in agriculture, fisheries and food* (Skinner, F. A. & Carr, J. G., eds). Soc. appl. Bact. Symp. Ser. No. 4. London and New York: Academic Press, p. 125.

HOBSON, P. N. & HOWARD, B. H. (1969). Microbial transformations. In *Handbook der Tierernährung*, **1**, 207. Hamburg: Paul Parcy.

HOBSON, P. N. & MANN, S. O. (1957). Some studies on the identification of rumen bacteria with fluorescent antibodies. *J. gen. Microbiol.*, **16**, 463.

HOBSON, P. N. & MANN, S. O. (1961). Experiments relating to the survival of bacteria introduced into the sheep rumen. *J. gen. Microbiol.*, **24**, i.

HOBSON, P. N. & MANN, S. O. (1970). Applications of the Coulter Counter in microbiology. In *Automation, mechanization and data handling in microbiology* (Baillie, A. & Gilbert, R. J., eds). London and New York: Academic Press.

HOBSON, P. N. & SHAW, B. G. (1974). The bacterial population of piggery waste anaerobic digesters. *Water Res.*, **8**, 507.

HOBSON, P. N. & SMITH, W. (1963). Continuous culture of rumen bacteria. *Nature, Lond.*, **200**, 607.

HOBSON, P. N. & STEWART, C. S. (1970). Growth of two rumen bacteria in mixed culture. *J. gen. Microbiol.*, **63**, 11.

HOBSON, P. N. & SUMMERS, R. (1966). Effect of growth rate on the lipase activity of a rumen bacterium. *Nature, Lond.*, **209**, 736.

HOBSON, P. N. & SUMMERS, R. (1967). The continuous culture of anaerobic bacteria. *J. gen. Microbiol.*, **47**, 53.

HOBSON, P. N. & SUMMERS, R. (1972). ATP pool and growth yield in *Selenomonas ruminantium*. *J. gen. Microbiol.*, **70**, 351.

HOBSON, P. N., MACKAY, E. S. M. & MANN, S. O. (1955). The use of fluorescent antibody in the identification of rumen bacteria. *Res. Corres.*, **8**, 30.

HOBSON, P. N., MANN, S. O. & OXFORD, A. E. (1958). Some studies on the occurrence and properties of a large Gram-negative coccus from the rumen. *J. gen. Microbiol*, **19**, 462.

HOBSON, P. N., MANN, S. O. & SMITH, W. (1962). Serological tests of a relationship between rumen selenomonads *in vitro* and *in vivo*. *J. gen. Microbiol.*, **29**, 265.

HOBSON, P. N., MCDOUGALL, E. I. & SUMMERS, R. (1968). The nitrogen sources of *Bacteroides amylophilus*. *J. gen. Microbiol.*, **50**, 1.

HOBSON, P. N., BOUSFIELD, S. & SUMMERS, R. (1974). The anaerobic digestion of organic matter. *Crit. Rev. Environ. Cont.*, **4**, 131.

HUNGATE, R. E. (1966). *The rumen and its microbes*. New York and London: Academic Press.

HUNGATE, R. E. (1969). A roll tube method for cultivation of strict anaerobes. In *Methods in microbiology* (Norris, J. R. & Ribbons, D. W., eds). Vol. 3B. New York and London: Academic Press, p. 117.

IANNOTTI, E. L., KAFKEWITZ, D., WOLIN, M. J. & BRYANT, M. P. (1973). Glucose fermentation products of *Ruminococcus albus* grown in continuous culture with *Vibrio succinogenes*; changes caused by interspecies transfer of H_2. *J. Bacteriol.*, **114**, 1231.

KURIHARA, Y., EADIE, J. M., HOBSON, P. N. & MANN, S. O. (1968). Relationship between bacteria and ciliate protozoa in the sheep rumen. *J. gen. Microbiol.*, **51**, 267.

LYSONS, R. J., ALEXANDER, T., HOBSON, P. N., MANN, S. O. & STEWART, C. S. (1971). Establishment of a limited rumen microflora in gnotobiotic lambs. *Res. Vet. Sci.*, **12**, 486.

LYSONS, R. J., ALEXANDER, T., HOBSON, P. N., MANN, S. O. & STEWART, C. S. (1976). Rumen bacteria in gnotobiotic lambs. *J. gen. Microbiol.*, **94**, 257.

MANN, S. O. & STEWART, C. S. (1974). Establishment of a limited rumen flora in gnotobiotic lambs fed on a roughage diet. *J. gen. Microbiol.*, **84**, 379.

MEERS, J. L. & TEMPEST, D. W. (1968). The influence of extracellular products on the behaviour of mixed microbial populations in magnesium-limited chemostat culture. *J. gen. Microbiol.*, **52**, 309.

PURDOM, M. R. (1963). Micromanipulation in the examination of rumen bacteria. *Nature, Lond.*, **198**, 307.

SHAW, B. G. (1971). *A practical and bacteriological study of the anaerobic digestion of waste from an intensive pig unit*. Ph.D. Thesis, Univ. of Aberdeen.

SHAPTON, D. A. & BOARD, R. G. (eds) (1971). *Isolation of anaerobes*. Soc. appl. Bact. Tech. Ser. No. 5. London and New York: Academic Press.

Methods for the Study of the Metabolism of Rumen Ciliate Protozoa and their Closely Associated Bacteria

G. S. Coleman

ARC Institute of Animal Physiology, Babraham, Cambridge, Cambridgeshire, England

Introduction

Rumen protozoa grow *in vivo* in a thick soup of bacteria and other particulate matter and can only be grown *in vitro* in a bacteriologically poor medium containing bacteria, starch grains and ground dried grass. Bacteria (up to 100 protozoon^{-1}) are present in vesicles in the endoplasm of most rumen ciliate protozoa (White, 1969) and up to 1000 protozoon^{-1} are also attached to the pellicle of *Epidinium* spp. (Coleman and Hall, 1974) and *Eudiplodinium* spp. Some of these bacteria have been cultured and there is evidence that all are metabolically active. In investigations on the metabolism of washed suspensions of these protozoa it is therefore necessary to distinguish between the products of bacterial and protozoal metabolism. If the source of protozoa is the rumen of a sheep containing only one species of ciliate protozoa, the method of Heald and Oxford (1953) without the addition of a sugar can be used successfully to remove plant material; the free bacteria and protozoa are then separated by differential centrifugation. However, although it is easy to separate small protozoa grown *in vitro* from plant material, it is impossible to separate a large protozoon such as *Polyplastron multivesiculatum* from the materials on which it was fed. As these plant materials are contaminated with bacteria it is important that their contribution to the metabolic activity of the suspension is recognized. The problem is further complicated by the ability of all protozoal species, but especially the *Entodinium* spp. (Coleman, 1964, 1972), to engulf bacteria and other particulate matter that is small enough to pass down the oesophagus.

General Methods

Source of protozoa

All protozoa were taken from the ovine rumen. *Entodinium caudatum* was isolated by enrichment and the other protozoa by the removal of single protozoa from crude rumen contents.

Culture of protozoa

Entodinium caudatum was grown as described by Coleman (1958, 1960, 1971) and *Epidinium ecaudatum caudatum* and *Polyplastron multivesiculatum* as described by Coleman (1971) and Coleman *et al.* (1972). All cultures, which contained bacteria, were fed with starch and ground dried grass each day.

Preparation of washed suspensions of protozoa

The protozoa were present as a loose pellet containing some grass, at the bottom of the culture tubes. After removal of the surface scum and most of the medium, the protozoa and the remaining medium were transferred to a 20 × 2·5 cm tube and allowed to stand until any grass present had sunk to the bottom, leaving the protozoa in the supernatant fluid. This supernatant fluid was transferred to centrifuge tubes, the residual grass washed with salt solution D* (Coleman, 1972) and the washings added to the supernatant fluid. The protozoa were spun down and washed four times in salt solution D, through which 95% N_2 + 5% CO_2 had been bubbled vigorously for 3 min, on a bucket-head centrifuge for 20 s from starting; the maximum speed was equivalent to 200 *g*. The protozoa were finally resuspended in salt solution D and 95% N_2 + 5% CO_2 bubbled through the suspension prior to addition to the incubation medium.

Incubation conditions

The medium consisted of 0·05–0·5 ml ^{14}C-labelled sugar or amino acid or ^{14}C-labelled bacteria, 0–0·05 ml antibiotic, 1·0 ml protozoal suspension and 0–0·45 ml salt solution D, to give a volume of 1·5 ml. Experiments were made in 80 × 13 mm thick-walled centrifuge tubes which, after inoculation, were gassed with 95% N_2 + 5% CO_2, sealed with a rubber bung and incubated at 39°. At the end of an experiment, the

* g litre^{-1}: K_2HPO_4 0·64; KH_2PO_4 0·55; NaCl 0·064; $CaCl_2$ (dried) 0·005; $MgSO_47H_2O$ 0·009.

culture was centrifuged and the protozoa washed three times at 200 *g* for 30 s in salt solution D. Under these conditions less than 0·5% of the bacterial ^{14}C appeared in the "protozoal fraction" in the absence of protozoa. Where the protozoa were sonicated before the incubation, the centrifugation at the end was at 7000 *g* for 20 min. The pellet was finally suspended in 2·0 ml water and a sample plated out for the estimation of ^{14}C (Coleman, 1969).

Breakage of protozoa

The protozoa in a suspension were broken by immersion of the tube to the depth of the liquid in the tube, in a 80 kc s^{-1} 40 W ultrasonic cleaning bath (KG 80/1, manufactured by Kerrys of Chester Hall Lane, Basildon, Essex) for 15 s. Unless the crude sonicate was required, the whole was centrifuged at 7000 *g* for 20 min. The supernatant fluid from this centrifugation is hereafter referred to as the "broken-cell supernatant fluid" and the pellet, after washing once in water, as the "broken-cell pellet". This latter fraction contained all the viable bacteria and the polysaccharide granules in the homogenate.

Bacteriological culture techniques

^{14}C-labelled *Escherichia coli* and *Klebsiella aerogenes* were grown at 39° in C medium (Roberts *et al.*, 1955) containing ^{14}C-glucose 2 mg ml^{-1} and (U-^{14}C) glucose 0·4 μCi ml^{-1} and aerated during growth by passing sterile air into the medium through a cottonwool-plugged Pasteur pipette. ^{14}C-labelled *Proteus mirabilis* was grown at 39° in a medium (designated YTG) that contained ($litre^{-1}$) salt solution D 500 ml; Difco yeast extract 2 g; Difco tryptose 2 g; glucose 2 g and (U-^{14}C) glucose 40 μCi, and which was aerated as described above.

Colony counts of aerobic bacteria were made by serial ten-fold dilutions in YTG medium followed by plating on the same medium plus 1% agar and incubation at 39°. Colony counts of typical anaerobic rumen bacteria were made on the medium (BR) of Bryant and Robinson (1961) as described by White (1969).

Number of Viable Bacteria inside *Entodinium caudatum*

Entodinium caudatum when freshly harvested from the growth medium, contains bacteria in vesicles in the endoplasm (Coleman and Hall, 1969). To determine the number of these that were viable, the protozoa were first washed several times on the centrifuge for 30 s from starting (maximum speed equivalent to 200 *g*) in sterile salt solution D to reduce the number of free bacteria to less than one protozoon^{-1}. The protozoa were then sonicated and samples plated out on BR and YTG media to determine the total number of viable bacteria present. In protozoa last fed 18 h before harvesting, the number of viable bacteria rose from 1 initially to 21·8 ± 8·2 protozoon^{-1} after 5 s sonication when 85% of the protozoa had been broken. Although there was 100% protozoal breakage after 10 s, the number of viable bacteria declined after 5 s sonication until after 1 min there were only 15 protozoon^{-1} (White, 1969). This suggests that the bacteria were killed slowly by the mild sonication conditions used and that the curve should be extrapolated back to zero time to give an initial count of 40 bacteria protozoon^{-1}. However, as there was considerable variation between different batches of protozoa, the number of bacteria was taken in practice as that found after 5 s sonication.

Approximately 90% of the bacteria isolated were vigorously growing Gram-negative rods identified as *Klebsiella aerogenes* and *Proteus mirabilis* which grew well on many media aerobically and anaerobically. The relative numbers of each were determined by the addition of chloramphenicol 50 μg ml^{-1} which specifically inhibited the growth of *Kl. aerogenes*. As colonies of any strictly anaerobic bacteria present in anaerobic cultures were rapidly overgrown by these two bacteria, their growth had to be inhibited by the addition of sodium azide 500 μg ml^{-1}. Under these conditions a number of typical rumen bacteria were isolated but their numbers never exceeded 2 protozoon^{-1}.

Engulfment, Killing and Digestion of Bacteria by Ciliate Protozoa

The engulfment of bacteria is studied by incubating together a washed suspension of the protozoa and a washed suspension of ^{14}C-labelled bacteria and measuring the incorporation of ^{14}C into the protozoal fraction. The bacteria were labelled by growth on medium in the presence of (U-^{14}C) glucose or, if the medium was a rich organic broth, in the presence of (8-^{14}C) guanine (Coleman, 1967). Mixed rumen bacteria were labelled by incubation for 16 h in 10 ml crude rumen contents (centrifuged to remove protozoa) with: salt solution D 10 ml; 5% cellobiose 1·0 ml; 5% NH_4Cl 0·5 ml; 1 mM Na_2SO_4 1 ml and ^{35}S-SO_4, 20 μCi ml^{-1}, 1 ml under 95% N_2 + 5% CO_2. The number of bacteria engulfed was calculated from the initial specific activity of the bacterial suspension and the amount of ^{14}C or ^{35}S found in the protozoal fraction after correction for "uptake" in the control incubations. To obtain meaningful results it was necessary to use other controls apart from that in the absence of protozoa and those used have been the incubation of labelled bacteria with boiled or sonicated protozoa at 39° and incubation with intact protozoa at 0°. Depending on the bacterium and protozoal species used, these can give higher or lower uptakes than those in the absence of protozoa, and boiled or sonicated protozoal preparations often gave higher uptakes initially, and after incubation, than those obtained with living protozoa. It is likely that the ^{14}C-labelled bacteria attach themselves more readily to the plant material and protozoal bodies than they do to the outside of living protozoa and untreated plant material. As non-specific attachment of the bacteria to the material in the suspension was the same at 0° as at 39°, incubation at 0° has been used as the control for most experiments. All incorporations were corrected for uptakes obtained initially on mixing the bacterial and protozoal suspensions together and separating them again as quickly as possible.

With those species that can be prepared free of plant material, e.g. *Ent. caudatum*, the survival of bacteria engulfed during an incubation with ^{14}C-labelled bacteria can be measured by making viable count determinations on washed protozoa before and after sonication, and comparing these with estimations of the number of bacteria engulfed obtained from ^{14}C measurements. For the viable counts it is of course necessary to correct the results for values obtained before incubation and for incubation in the absence of bacteria.

The digestion of bacteria is measured by the appearance of ^{14}C from ^{14}C-labelled bacteria in the protozoal broken-cell supernatant fluid or the supernatant fluid obtained after removal of the bacteria by centrifugation at 4000 *g* for 20 min from incubation medium. Unfortunately incubation at 0° is a poor control for the non-specific release of ^{14}C and incubation in the absence of any protozoal preparation must be used. In the presence of boiled or sonicated protozoa there was often a larger release of ^{14}C into the medium than in the presence of live protozoa, so these too could not be used as controls.

Measurement of Uptake of Soluble Compounds by Ciliate Protozoa

Effect of protozoal breakage and of antibiotics

As shown above, gentle sonication of protozoal suspensions can be used to break the protozoa without damaging the bacteria. If it is assumed that the protozoal fragments have no ability to incorporate ^{14}C-labelled glucose or amino acids, any incorporation into the sonicate must be into the bacteria present. This assumption is supported by the inhibitory action of antibiotics (e.g. chloramphenicol and ampicillin) on this incorporation (Figs 1, 2). Incubation of intact protozoa in the presence of

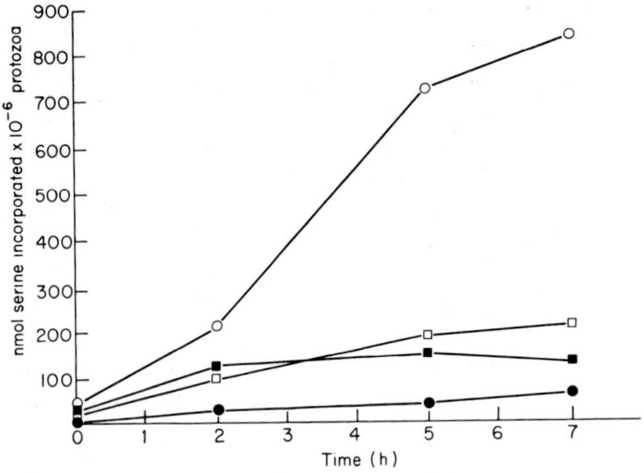

FIG 1. Effect of presence (●, ■) or absence (○, □) of ampicillin (1 mg ml^{-1}) on the incorporation of ^{14}C from 0·1 mM (U-^{14}C) serine into the protozoal broken-cell pellet of *Epidinium ecaudatum caudatum*. □, ■, intact protozoa incubated with amino acid and then washed, sonicated and centrifuged to obtain pellet fraction; ○, ●, sonicated protozoa incubated with amino acid and then centrifuged at 7000 *g* and the pellet obtained washed twice under the same conditions. This pellet was the same as that obtained from intact protozoa except that incubation with ^{14}C-labelled serine occurred after, rather than before, sonication.

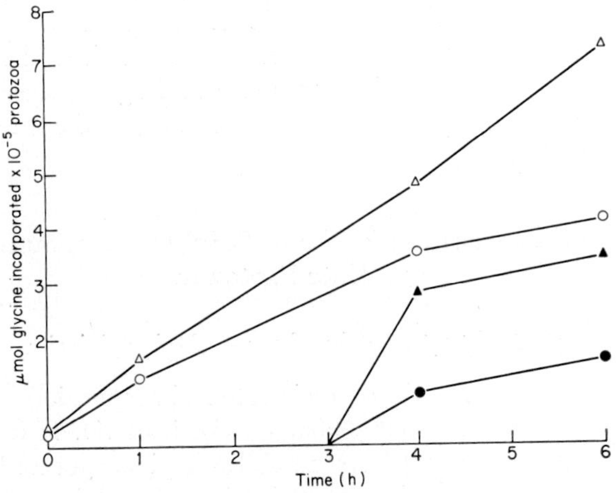

FIG. 2. Effect of presence (○, ●) or absence (△, ▲) of chloramphenicol (80 μg ml^{-1}) on the incorporation of ^{14}C from (U-^{14}C) glycine by *Epidinium ecaudatum caudatum*. ●, ▲ glycine added at 3 h to protozoa that were incubated throughout in the presence or absence of chloramphenicol.

antibiotic might therefore be expected to prevent incorporation of a ^{14}C-labelled amino acid by the bacteria and enable uptake into the protozoa to be measured. The addition of ampicillin 1 mg ml^{-1} decreased incorporation of ^{14}C from ^{14}C-labelled serine by *Epi. ecaudatum caudatum* into the broken-cell pellet fraction of intact protozoa (Fig. 1) but had no effect on uptake into the broken-cell supernatant fluid. As this latter fraction is of protozoal origin whereas the pellet fraction contains bacteria and protozoal material, this shows that ampicillin had no effect on uptake of serine by the protozoon itself. Unfortunately antibiotics penetrate only slowly into these ciliate protozoa (Coleman, 1962), and the bacteria may not be inhibited during the course of the experiment, especially with an antibiotic such as ampicillin that only acts on growing bacteria. This can be overcome in part by incubation of the protozoa with the antibiotic for 3 h before addition of the ^{14}C-labelled compound (Fig. 2). In an experiment on the uptake of ^{14}C-labelled glycine over 1 h by *Epi. ecaudatum caudatum* in the presence of chloramphenicol 80 $\mu g\ ml^{-1}$, pre-incubation of the protozoa with the antibiotic for 3 h decreased incorporation of ^{14}C by 31%, whereas in the corresponding experiment, made in the complete absence of antibiotic, incorporation increased by 86%. This suggests that without antibiotic the number of bacteria associated with the protozoa increased and that in the presence of chloramphenicol uptake by the bacteria decreased. However, it is possible for the addition of antibiotic to inhibit the uptake of soluble compounds by the protozoon itself as was shown for the inhibition by penicillin of glucose uptake by *Ent. caudatum* (Coleman, 1969). In this respect each combination of substrate, antibiotic and protozoal species may behave differently. Attempts to prepare bacteria-free suspensions of *Entodinium caudatum* by incubation with penicillin, neomycin, streptomycin and dihydrostreptomycin were only partially successful because, although the number of bacteria was reduced to less than one per protozoon, the resultant protozoa were moribund and metabolically inactive (Coleman, 1962).

Measurement in the presence of plant material

Although it is not possible to separate *Polyplastron multivesiculatum* grown *in vitro* from plant material in the culture, the converse separation was possible. An equal volume of "plant material" to the protozoa-rich fraction was therefore used as a control and both fractions incubated with and without chloramphenicol 80 $\mu g\ ml^{-1}$. In the absence of antibiotic, uptake of ^{14}C from ^{14}C-glycine by the protozoal fraction was linear for 6 h whereas the rate of uptake into the plant material decreased

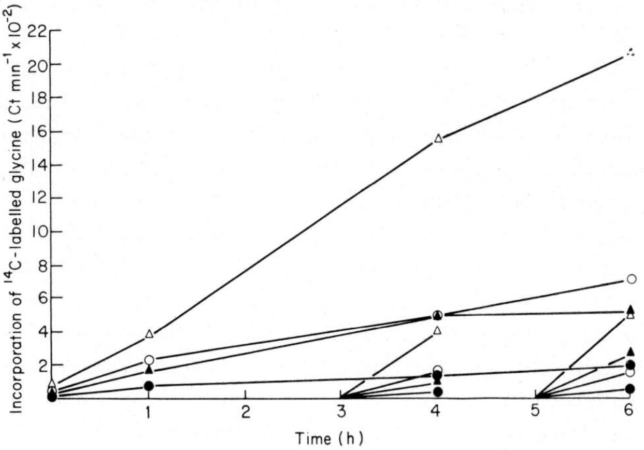

FIG. 3. Effect of presence (o, ●) or absence (△, ▲) of chloramphenicol (80 μg ml^{-1}) on the incorporation of ^{14}C from (U-^{14}C) glycine, added initially or after 3 or 5 h, by protozoa plus plant material (o, △) or an equal volume of plant material alone (●, ▲).

with time (Fig. 3). The addition of chloramphenicol decreased incorporations into the protozoa and plant fractions by 57 and 66% respectively. Preincubation of either fraction for 3 or 5 h in the absence of antibiotic before incubation with the ^{14}C-glycine for 1 h, increased the uptake of ^{14}C presumably due to the increase in the number of bacteria present. In contrast, preincubation with chloramphenicol for 3 or 5 h decreased uptake into the protozoal fraction to the same extent, suggesting that bacterial metabolism was inhibited after 3 h and that only protozoal metabolism was measured thereafter. Unfortunately, there was still some incorporation of ^{14}C into the plant material even after 5 h incubation with chloramphenicol, although this could be adsorption rather than uptake by bacteria.

Autoradiography in the Electron Microscope

The problems of whether incorporation of ^{14}C from a ^{14}C-labelled substrate is into bacteria or protozoal material, or whether bacteria found in the protozoal fraction are inside the protozoa or attached to the pellicle, can be answered qualitatively by autoradiography in the electron microscope of protozoa that have metabolized tritium-labelled materials (Coleman and Hall, 1974). The organelles into which the tritium was incorporated can be clearly seen by the presence of overlying silver grains, but as even the smallest rumen ciliate protozoon is 40 μm long,

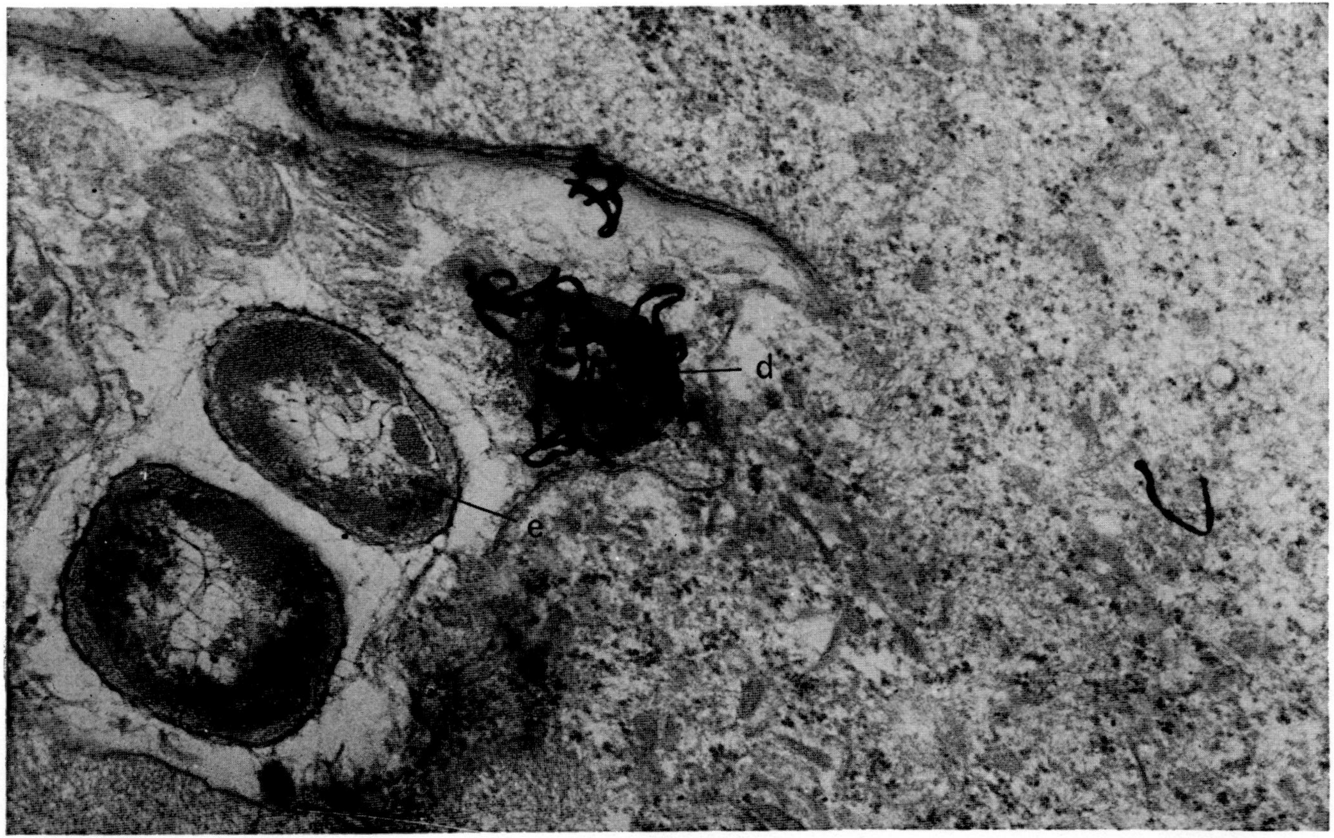

Figure 4

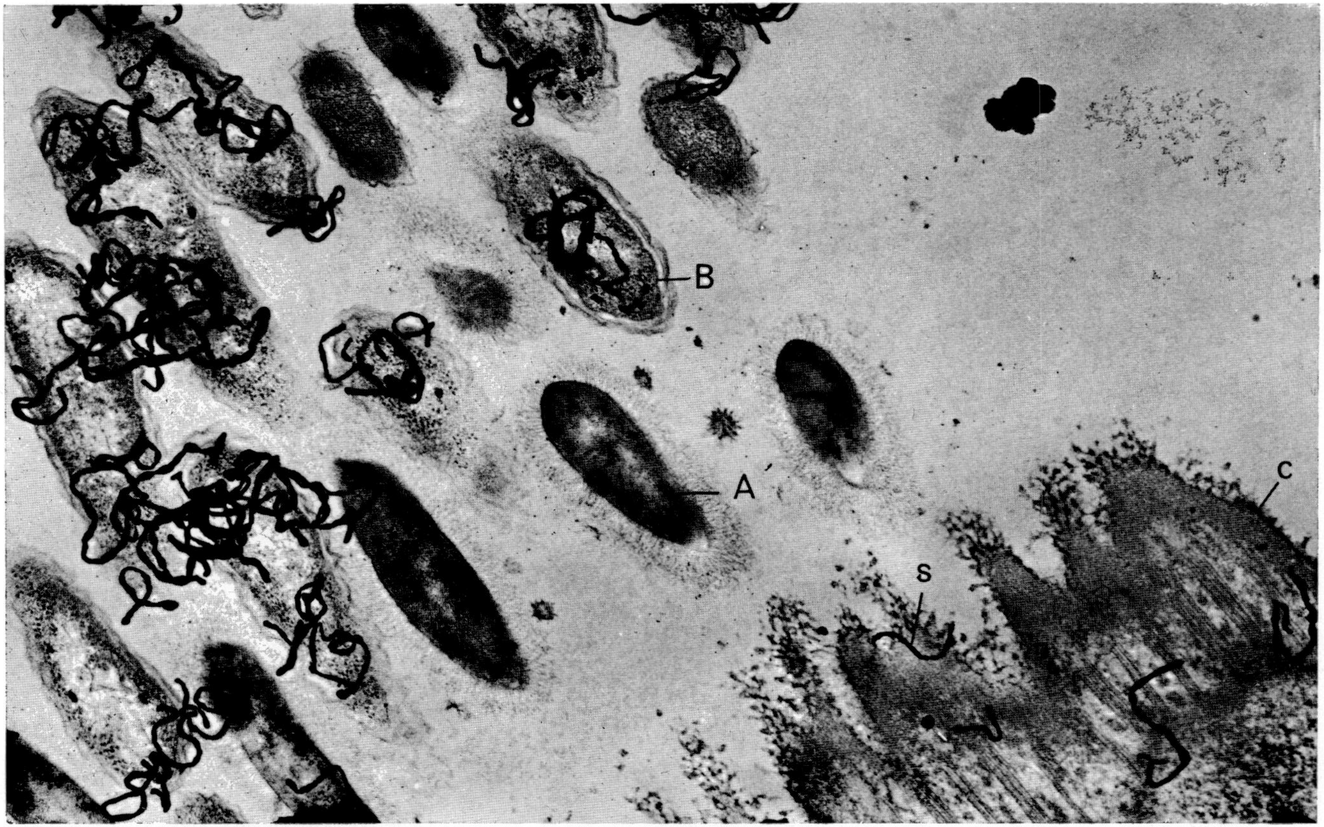

Figure 5

it is rarely possible to see a complete section on a grid. Any quantitative studies of the relative incorporation of tracer into different structures must therefore be open to large errors. Furthermore, different protozoa in the same suspension take up different amounts of tracer and there are even different amounts of tracer in the same "organelle" in different parts of one protozoon (Figs 4, 5). The technique has therefore just been used to obtain information on whether incorporation of, e.g. ^{3}H-labelled glucose into bacteria or polysaccharide granules occurs under certain conditions. Quantitative data has been obtained by other methods.

Uptake of ^{3}H-labelled glycine by Epidinium ecaudatum caudatum

After incubation of a suspension of *Epi. ecaudatum caudatum* for 2 h with 100 μCi (2-^{3}H) glycine (29 mC mg^{-1}) there was extensive labelling of the protozoa. Tracer was found widely distributed through the cytoplasm and pellicle but none was present in the polysaccharide granules (Fig. 6). Some of the bacteria in the cytoplasm (Fig. 4) were labelled as were some of those attached to the pellicle (Figs 4, 6); the bacteria designated B (Coleman and Hall, 1974) always contained more tracer than those designated A. On addition of ampicillin 1 mg ml^{-1} and incubation for 5 h, tracer was again distributed through the cytoplasm and pellicle as shown in Fig. 6, but the type B bacteria attached to the pellicle contained proportionally less tracer than before and showed signs of lysis (Fig. 7).

Uptake of ^{3}H-labelled glucose by Epidinium ecaudatum caudatum

Autoradiography showed that on incubation of *Epi. ecaudatum caudatum* with (6-^{3}H) glucose, tritium was incorporated into the protozoal polysaccharide granules in the skeletal plates and free in the cytoplasm, but to a much smaller extent, especially at high glucose concentrations, into the bacteria attached to the pellicle (Coleman and Hall, 1974).

FIGS 4, 5 and 6. Electron micrographs of sections through *Epidinium ecaudatum caudatum* that had been incubated with (2–^{3}H) glycine for 2 h. Note that tracer is found distributed through the cytoplasm and that some bacteria attached to the pellicle (usually type B) are heavily labelled whereas others (usually type A) are not. Similarly, as shown in Fig. 4, some intracellular bacteria (d) are heavily labelled whereas others (e) are not. A, type A bacterium; B, type B bacterium; p, protozoal polysaccharide granule; c, pellicle of protozoon; s, silver grain. FIG. 4 × 48 000; FIG. 5 × 36 000; FIG. 6 × 24 000.

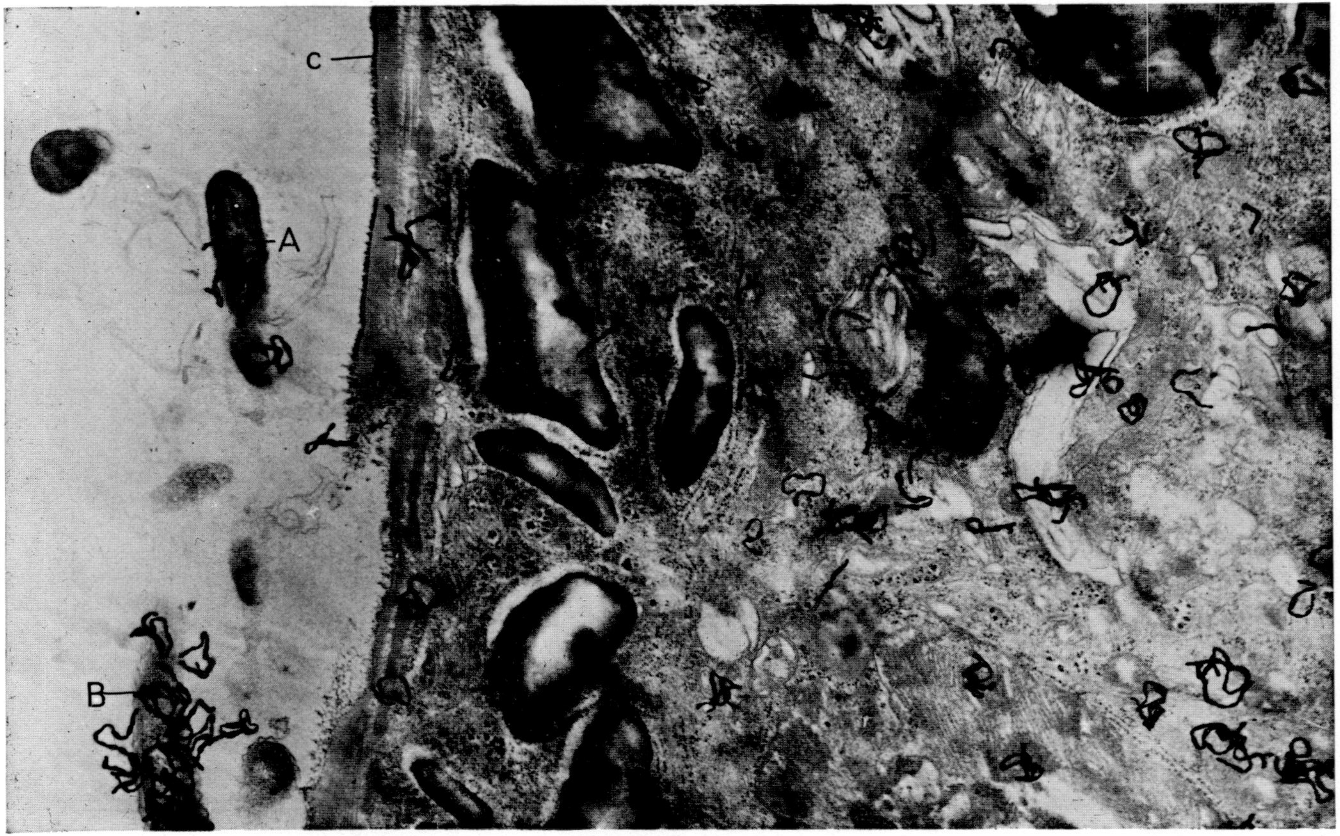

Figure 6

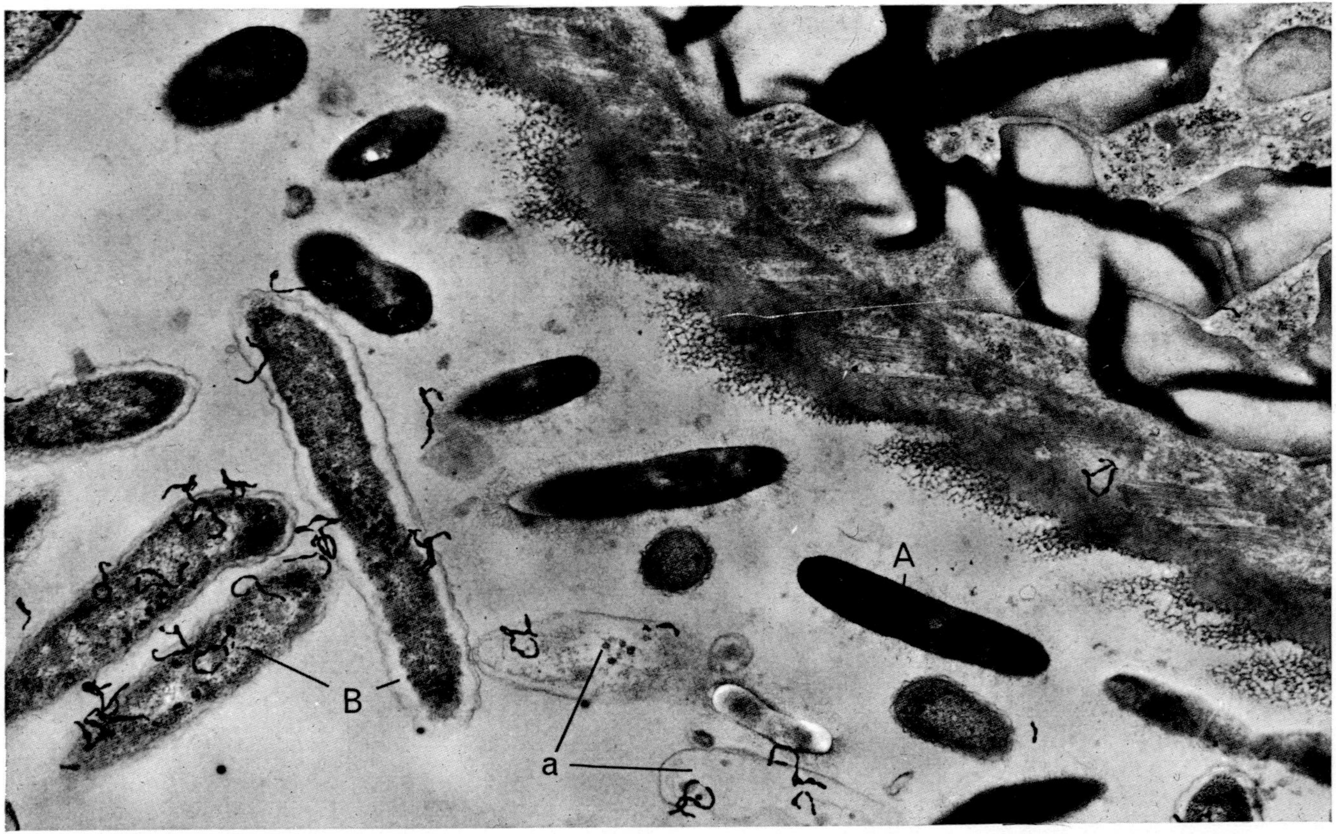

FIG. 7. Electron micrograph of a section through *Epidinium ecaudatum caudatum* that had been incubated with (2–^{3}H) glycine and ampicillin 1 mg ml^{-1} for 2 h. Note that some B type bacteria (a) have been lysed but still contain tracer. × 24 000.

Uptake of bacteria by Epidinium ecaudatum caudatum

By use of the method described above, Coleman and Laurie (1974) showed that *Epi. ecaudatum caudatum* took up *Pr. mirabilis* readily and it was assumed that the bacteria were engulfed into vesicles in the cytoplasm. However, it was possible that these bacteria were becoming attached to the outside of the pellicle next to the A and B type bacteria. Protozoal suspensions were therefore incubated with ^{3}H-labelled *Pr. mirabilis* and Fig. 8 shows that, after 1 h, the protozoa had in them large vesicles containing labelled bacteria and membraneous material. After 5 h incubation many labelled multimembrane systems, which must have been derived from the bacteria, were present in the cytoplasm but few vesicles or intact bacteria were apparent (Fig. 9). At neither time were any labelled bacteria attached to the pellicle. These results show that the correct interpretation had been placed on those obtained previously.

Uptake of glucose by Entodinium caudatum

Entodinium caudatum has no bacteria attached to its pellicle (Coleman and Hall, 1969), but contains in its cytoplasm bacteria that incorporate ^{14}C from ^{14}C-labelled glucose (see above). Autoradiography in the electron microscope of sections of protozoa that had metabolized (6-^{3}H) glucose for 5 h showed that tracer was taken up by the bacteria and was also present in the protozoal polysaccharide granules. In addition, some labelled, but partially digested, bacteria were present. This shows that bacteria which were alive at the beginning of the experiment had been subsequently killed and digested by the protozoon and provides evidence that the bacteria living in the cytoplasm are being continually digested by the host.

Biochemical Methods for Investigating the Uptake of Glucose by *Entodinium caudatum* and *Epidinium ecaudatum caudatum*

After gentle sonication of suspensions of *Ent. caudatum* or *Epi. ecaudatum caudatum* to break the protozoa, but not the bacteria or polysaccharide granules, and centrifugation of the sonicate at 7000 *g* for 20 min, both of these bodies appeared in the pellet. The problem was to find a method of separating them. However, it was first necessary to find a way of labelling only one of these and the easiest method was to label the bacteria by incubation of sonicated protozoa with ^{14}C-labelled glucose under which conditions most of the ^{14}C was in the bacterial capsular polysaccharide.

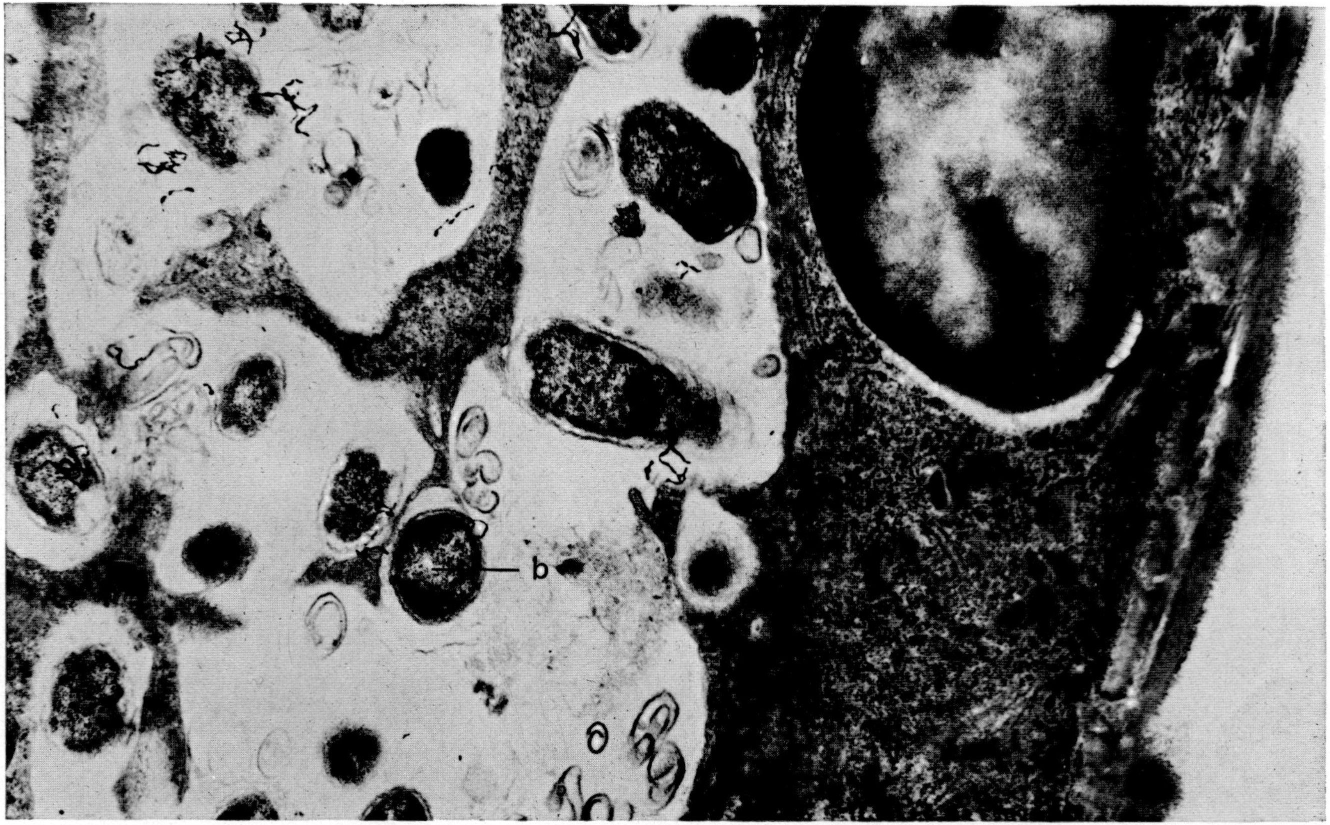

FIG. 8. Electron micrograph of a section through *Epidinium ecaudatum caudatum* that had been incubated with ^{3}H-*Proteus mirabilis* for 1 h. Note the presence of labelled bacteria (b) in a vesicle in the endoplasm. × 21 000.

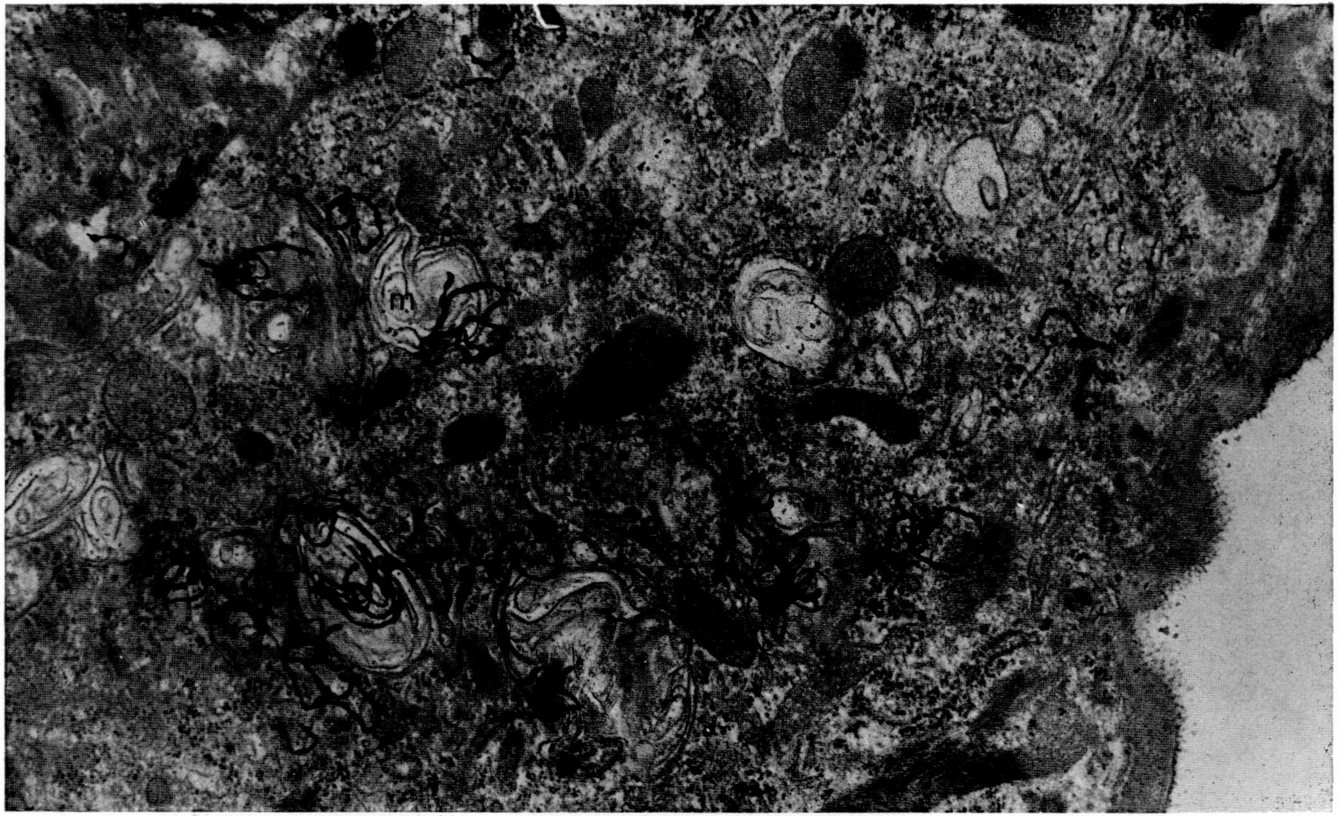

FIG. 9. Electron micrograph of a section through *Epidinium ecaudatum caudatum* that had been incubated with ^{3}H-*Proteus mirabilis* for 5 h. Note the presence of many labelled multimembrane systems (m) in the cytoplasm. $\times$ 36 000.

With *Ent. caudatum* the washed pellet from this incubation was resuspended in water and samples layered on top of 2 ml quantities of sucrose of differing molarities and centrifuged at 1900 *g* for 10 min in a bucket-head centrifuge. The amount of ^{14}C and the number of protozoal polysaccharide granules were then estimated in the washed pellets from this treatment. Figure 10 shows that the amount of ^{14}C (i.e. the number of bacteria) in the pellet declined rapidly with increasing sucrose concentration and that with 2 M sucrose only 2·5% of the ^{14}C was found in

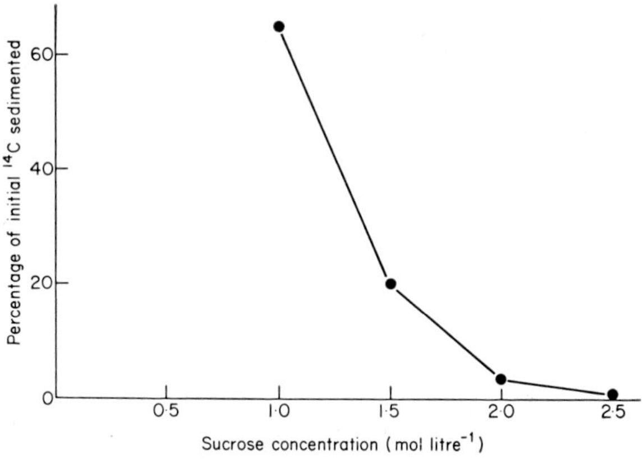

FIG. 10. Effect of sucrose concentration on the sedimentation of ^{14}C-labelled bacteria (in a gently sonicated preparation of *Entodinium caudatum*) through 2 ml sucrose contained in a 13 mm diameter tube and centrifuged for 10 min at 1900 g.

the pellet. In contrast 30% of the original number of polysaccharide granules were present and it was therefore possible to obtain an estimate of the amount of ^{14}C incorporated by intact protozoa into these granules by gentle sonication of the protozoa and centrifugation of the resultant broken-cell pellet through 2 M sucrose and multiplying the amount of ^{14}C in the washed pellet from this treatment by 3·3. For the purpose of this calculation the 2·5% of the bacterial ^{14}C that centrifuged through sucrose was ignored. The amount of ^{14}C in the intracellular bacteria was the difference between the total ^{14}C in the broken-cell pellet and the ^{14}C in the protozoal polysaccharide measured as described above. This method has been used to determine the effect of concentration of ^{14}C-labelled glucose on the incorporation of ^{14}C into the different cell fractions (Coleman, 1969).

Unfortunately this method did not work well with *Entodinium simplex* (Coleman, 1972) or *Epi. ecaudatum caudatum* (Coleman and Laurie, 1974) and with the latter protozoon the following method was used. The broken-cell pellet was treated with a Mullard MSE ultrasonic drill (type 7685/2) with a 20 mm diameter probe just touching the surface of 2 ml suspension in a 50 ml beaker surrounded by ice. The mixture was then centrifuged and measurements made of the amount of ^{14}C and the number of polysaccharide granules in the washed pellet. To determine the minimum time of sonication required to break up the bacteria, broken-cell pellet containing only labelled bacteria (prepared as described above) was treated for various times and the amount of ^{14}C in the final pellet measured. Figure 11 shows that after 5 min sonication only 6% of

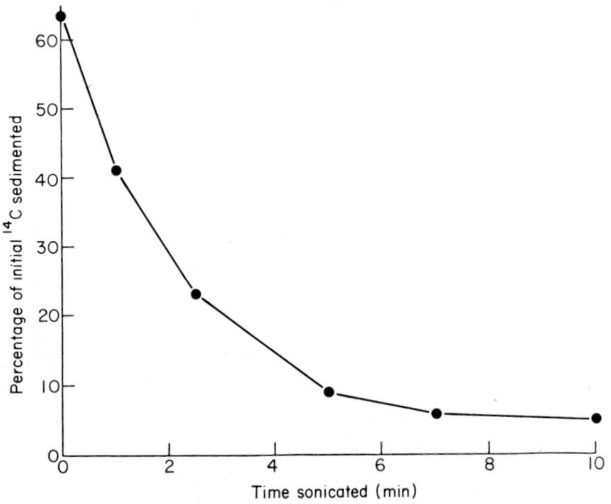

FIG. 11. Effect of time of sonication with an ultrasonic drill of ^{14}C-labelled bacteria (in a gently sonicated preparation of *Epidinium ecaudatum caudatum*) on the rate at which ^{14}C was sedimented subsequently on centrifugation at 1900 g for 10 min.

the original ^{14}C was present in the pellet and this time was adopted. The polysaccharide granules were unaffected by this treatment. Any appreciable amount of ^{14}C found in the washed pellet after sonication of the broken-cell pellet from intact protozoa that had been incubated with ^{14}C-labelled glucose was assumed to be in protozoal polysaccharide. For example, in the presence or absence of ampicillin 1 mg ml^{-1}, 56 and 38% of the glucose polysaccharide found in the protozoa from ^{14}C-labelled glucose was in the form of protozoal polysaccharide granules, the remainder presumably being bacterial capsular polysaccharide (Coleman and Laurie, 1974).

Differential Labelling of Bacteria used as Food for *Entodinium caudatum*

When the two bacteria normally found in the cultured form of *Ent. caudatum*, i.e. *Kl. aerogenes* and *Pr. mirabilis*, are grown in conventional bacteriological medium with forced aeration and then incubated with the protozoa, both are engulfed readily but *Pr. mirabilis* is killed and digested only slowly (Coleman, 1967). In contrast *Kl. aerogenes* is digested comparatively rapidly, 65% of the bacteria taken up during $2\frac{1}{2}$ h of continuous engulfment being dead at the end. The survival time of the bacteria inside the protozoa was therefore possibly only 1 h. In freshly harvested protozoa, *Pr. mirabilis* might therefore be expected to be the predominant bacterium, but the converse was found. When these freshly harvested protozoa were incubated in sterile salts medium, the number of viable bacteria in each protozoon and in each ml of medium was, respectively, 3 and 2×10^6 initially, 1 and 10×10^6 after 5 h and 102 and 760×10^6 after 24 h showing that the protozoa were not killing the bacteria faster than they could grow. As over 50% of the bacteria were *Kl. aerogenes*, these results suggest that this bacterium was, in the culture and protozoa, comparatively resistant to digestion. An explanation was therefore sought for the different rates at which this bacterium was killed under different conditions.

As *Kl. aerogenes* has, when grown anaerobically on glucose-containing media, a capsule composed of glucose polysaccharide and as this is absent in bacteria grown with forced aeration, the presence or absence of this capsule may determine the rate at which the bacteria are killed and digested. To test this theory *Kl. aerogenes* was prepared with and without a capsule and labelled in different parts of the cell. These bacteria were then incubated with washed suspensions of *Ent. caudatum* and the appear-

TABLE 1. The digestion by *Entodinium caudatum* of *Klebsiella aerogenes* labelled in different parts of the cell

Conditions for labelling bacteria	Capsule	Rest of cell	Bacteria digested protozoon^{-1} in 3h
A. Incubation anaerobically of washed suspension with (U^{-14} C) glucose for 30 min	Labelled	Unlabelled	25
B. Growth with forced aeration in presence of (U-^{14} C) glycine	Absent	Labelled	252
C. Growth anaerobically in presence of (U-^{14} C) glycine	Present, unlabelled	Labelled	152

ance of ^{14}C in the medium measured (Table 1). The results show that capsular material was digested comparatively slowly (culture A) and that the remainder of the bacterial cell was digested more rapidly when the capsule was absent (culture B) than when it was present (culture C) (Coleman, 1975). This is taken as evidence that *Kl. aerogenes* survives inside the protozoon because it synthesises a capsule, under the anaerobic conditions prevailing there, from the glucose in the protozoal pool (Coleman, 1969).

If the protozoa were last fed with starch 18, 42 or 66 h before harvesting, the number of viable bacteria present per protozoon was 37, 10 and 2 respectively. As the amount of free glucose in the protozoon also decreases with time since the protozoon was last fed (Coleman, 1969), this is taken as further evidence that the bacteria are progressively less able to withstand the killing action of protozoal enzymes as the substrate for synthesis of the capsule disappears.

Acknowledgement

I wish to thank Mr. F. J. Hall for taking and providing the electron micrographs.

References

BRYANT, M. P. & ROBINSON, I. M. (1961). An improved nonselective culture medium for ruminal bacteria and its use in determining diurnal variation in numbers of bacteria in the rumen. *J. Dairy Sci.*, **44**, 1446.

COLEMAN, G. S. (1958). Maintenance of oligotrich protozoa from the sheep rumen *in vitro*. *Nature, Lond.*, **182**, 1104.

COLEMAN, G. S. (1960). The cultivation of sheep rumen oligotrich protozoa *in vitro*. *J. gen. Microbiol.*, **22**, 555.

COLEMAN, G. S. (1962). The preparation and survival of almost bacteria-free suspensions of *Entodinium caudatum*. *J. gen. Microbiol.*, **28**, 271.

COLEMAN, G. S. (1964). The metabolism of *Escherichia coli* and other bacteria by the rumen ciliate *Entodinium caudatum*. *J. gen. Microbiol.*, **37**, 209.

COLEMAN, G. S. (1967). The metabolism of the amino acids of *Escherichia coli* and other bacteria by the rumen ciliate *Entodinium caudatum*. *J. gen. Microbiol.*, **47**, 449.

COLEMAN, G. S. (1969). The metabolism of starch, maltose, glucose and some other sugars by the rumen ciliate *Entodinium caudatum*. *J. gen. Microbiol.*, **57**, 303.

COLEMAN, G. S. (1971). The cultivation of rumen Entodiniomorphid protozoa. In *Isolation of anaerobes* (Shapton, D. A. & Board, R. G., eds). Soc. appl. Bact. Tech. Ser. No. 5. London and New York: Academic Press, p. 159.

COLEMAN, G. S. (1972). The metabolism of starch, glucose, amino acids, purines, pyrimidines and bacteria by the rumen ciliate *Entodinium simplex*. *J. gen. Microbiol.*, **71**, 117.

COLEMAN, G. S. (1975). The role of bacteria in the metabolism of rumen Entodiniomorphid protozoa. *Symp. Soc. Exptl. Biol.*, **29**, 533.

COLEMAN, G. S. & HALL, F. J. (1969). Electron microscopy of the rumen ciliate *Entodinium caudatum* with special reference to the engulfment of bacteria and other particulate matter. *Tissue and Cell*, **1**, 607.

COLEMAN, G. S. & HALL, F. J. (1974). The metabolism of *Epidinium ecaudatum caudatum* and *Entodinium caudatum* as shown by autoradiography in the electron microscope. *J. gen. Microbiol.*, **85**, 265.

COLEMAN, G. S. & LAURIE, J. I. (1974). The metabolism of starch, glucose, amino acids, purines, pyrimidines and bacteria by three *Epidinium* spp. isolated from the rumen. *J. gen. Microbiol.*, **85**, 244.

COLEMAN, G. S., DAVIES, J. I. & CASH, M. A. (1972). The cultivation of the rumen ciliates *Epidinium ecaudatum caudatum* and *Polyplastron multivesiculatum in vitro*. *J. gen. Microbiol.*, **73**, 509.

HEALD, P. J. & OXFORD, A. E. (1953). Fermentation of soluble sugars by anaerobic holotrich ciliate protozoa of the genera *Isotricha* and *Dasytricha*. *Biochem. J.*, **53**, 506.

ROBERTS, R. B., ABELSON, P. H., COWIE, D. E., BOLTON, E. T. & BRITTEN, R. J. (1955). Studies on biosynthesis in *Escherichia coli*. *Publs. Carnegie Instn.*, No. 607.

WHITE, R. W. (1969). Viable bacteria inside the rumen ciliate *Entodinium caudatum*. *J. gen. Microbiol.*, **56**, 403.

The Use of Continuous Cultures and Electronic Sizing Devices to Study the Growth of Two Species of Ciliated Protozoa

C. R. Curds, D. McL. Roberts and Chih-Hua Wu

British Museum (Natural History), Cromwell Road, London, England

Introduction

Although a considerable amount of effort has been spent studying the growth of single species of ciliated protozoa in pure cultures, little attention has been paid to the growth of mixed populations of these organisms. The obvious reasons for this lack of work are that methods have been either unavailable or not sufficiently well developed to enable the evaluation of the biomass of the constituent organisms. However, the continuing development of the Coulter Counter and associated equipment now makes it possible to distinguish between cells of different sizes, to estimate their numbers and to evaluate their mean cell volumes.

In the past decade some attention has been given to the quantitative aspects of the feeding and growth of ciliated protozoa upon bacteria, and several authors have turned to the use of continuous-culture techniques for these studies (Curds and Cockburn, 1971; Hamilton and Preslan, 1970; Jost *et al.*, 1973; Proper and Garver, 1966). Quantitative information obtained in this way makes it possible to estimate the magnitude of the effects of these predatory activities upon the bacterial flora in the aquatic environment. There are even fewer data available on the quantitative aspects of the predation of ciliates on organisms other than bacteria and the reader is recommended to the review by Curds and Bazin (1977) for an account of protozoan predation in batch and continuous cultures.

The purpose of the present paper is to record some of the methods which are being developed to study the predation of one ciliate species upon another. This forms part of a long-term project concerning the population dynamics and feeding kinetics of ciliated protozoa.

The Organisms

The ciliate *Tetrahymena pyriformis* has featured in a large proportion of the work carried out on the growth and biochemistry of ciliated protozoa (Elliott, 1973; Hill, 1972). This is prinicipally because of the ease with which it may be cultured in the absence of bacteria (axenic)—indeed it was the first of all animals to be cultivated under bacteria-free conditions. *Tetrahymena pyriformis* GL (NERC Culture Centre of Algae and Protozoa, 36 Storey's Way, Cambridge: strain L 1630/1 GL) was chosen as the prey for several reasons. First, a great deal of information on its physiology and nutrition is available; secondly, some work on its feeding kinetics has been published and thirdly, as it is an amicronucleate strain, it must rely solely upon asexual binary fission as a means of reproduction. It was routinely maintained axenically in plugged test tubes containing 10 g $litre^{-1}$ proteose peptone, 2·5 g $litre^{-1}$ yeast extract medium.

Several ciliates will feed upon *T. pyriformis* and these include the suctorian *Tokophrya infusionem* and the two holotrichs *Tetrahymena vorax* and *Tetrahymena patula*. The suctorian was found to have several disadvantages, the major ones being that it was difficult to maintain and that it has a sedentary mode of life. The tendency for an organism to settle on surfaces inside a chemostat vessel is an undesirable property. The two carnivorous holotrichs *T. vorax* and *T. patula* have a free-swimming habit and both are more easily cultivated than the suctorian. However, both *T. vorax* and *T. patula* have polymorphic life cycles which are linked to diet. When these organisms are fed upon small particulate food such as bacteria, the microstome (small-mouthed) form is adopted. In the presence of larger prey, such as *T. pyriformis*, microstomes transform into macrostome (large-mouthed) individuals. *Tetrahymena patula* was finally chosen as the experimental predator since although it does exist in both microstome and macrostome states it has a tendency to favour the macrostome form while *T. vorax* tends to remain as a microstome. Furthermore, the size difference between the predator and its prey is greater in the case of *T. patula*.

The life cycle of *T. patula* has a non-motile stage which is said by Williams (1960) to be a reproductive cyst. Although we have noted clumped and solitary rounded-up forms we have not yet been able to excyst them nor have we been able to find any evidence of the "invisible gelatinous envelope" reported by Williams (1960).

In continuous cultures of *T. patula* we therefore have the possibility of encountering all three morphological states of this organism together with its prey, *T. pyriformis*. In practice the encysted form is rare and usually can be ignored as it seems to appear only when the culture becomes contaminated. Since *T. patula* macrostomes are known to feed upon *T. patula* microstomes it is important to estimate and distinguish the individual biomass of the three remaining constituent populations. It is possible to do this with the aid of a Coulter Counter and Channelyzer. Suitable mathematical routines that can be used when there is an overlap in the size distributions of the individual populations are being investigated and these will be mentioned later.

Tetrahymena patula (NERC Culture Centre for Algae and Protozoa, strain L 1630/2 L–FF) was routinely maintained axenically in plugged test tubes containing 10 g $litre^{-1}$ proteose peptone, 2·5 g $litre^{-1}$ yeast extract as the growth medium. Organisms cultivated in this way initially grow as microstomes but as the soluble nutrients of the medium are depleted the population changes to macrostome organisms which prey upon the remaining microstome population. When larger populations were required for experimental purposes, Loefer's medium (Loefer *et al.*, 1958) was used as this medium gave a higher yield.

Continuous-culture Methods

The theory, use and advantages of continuous-culture methods are well known and several books on the subject have been published (Kubitschek, 1970; Malék and Fencl, 1966; Pirt, 1975). However, when one is dealing with a mixed population of organisms it may be necessary to adapt the chemostat system in order to obtain the required data. In the case of the continuous culture of a predator and its prey, two options are available. If the dynamic behaviour between the predator and its prey is required then a conventional single-stage chemostat system may be used. In that system both the predator and prey are actively growing together in the reactor and the populations may undergo some oscillatory behaviour. If, on the other hand, data on the feeding kinetics of the predator growing under steady-state conditions are required then a constant population of non-growing prey organisms must be supplied to the reactor containing the predator. This can be arranged by batch methods as carried out by Hamilton and Preslan (1970), although we prefer to do this continuously by the use of a two-stage system (Fig. 1). The first-stage reactor is used to convert the soluble components of the sterile growth medium into prey organisms. Some of the prey was continually wasted, while some was pumped to the second-stage reactor in which the predators were grown. The design essentially follows that of Curds and Cockburn (1971) except that a variable-speed peristaltic pump (Varioperpex 12 000 pump, LKB Instruments Ltd, 232 Addington Rd,

Croydon CR2 8YD) was used between the two reactors; this enabled the dilution rate of the second stage to be controlled by the flow rate. *Tetrahymena patula* was found to be a highly fragile organism that disrupted when stirred and aerated too vigorously. The effects were minimized by stirring the contents of the second reactor at a rate of no more than 20 r/min and by aerating very gently. The temperature of the cultures was kept constant at 25° by the use of thermostatically controlled water baths.

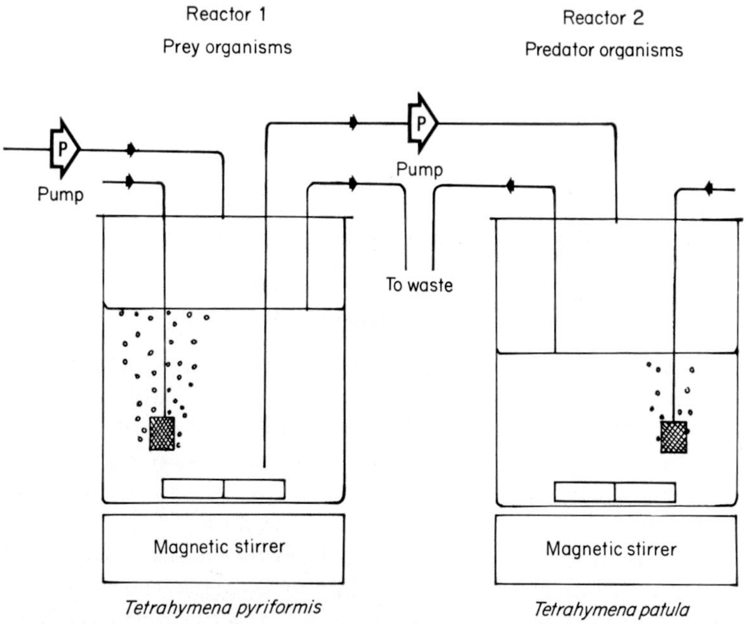

FIG. 1. Diagrammatic representation of a two-stage continuous-culture apparatus.

Using this apparatus *T. pyriformis* was grown in the first stage on a medium composed of 500 mg $litre^{-1}$ proteose peptone, 125 mg $litre^{-1}$ yeast extract and 100 mg $litre^{-1}$ glucose. Batch experiments have indicated that this medium is likely to be carbon limited. There is a linear response between glucose concentration and *T. pyriformis* population up to a concentration of at least 600 mg $litre^{-1}$ glucose. Similarly the addition of glucose stimulates the specific growth rate of *T. pyriformis*.

Beginning a continuous culture of *T. patula* can be a difficult procedure but the problems can be overcome by the adaptation of the *T. patula* to a diet of *T. pyriformis* before inoculation. When the two organisms are grown together in the presence of a high concentration of organic matter they compete (*T. patula* as a microstome) and under these conditions *T. pyriformis* is invariably the most successful. For these reasons the adaptation of *T. patula* to a diet of *T. pyriformis* is carried out in an environment lacking in soluble organic substrates. An axenic test-tube culture of each organism was washed and resuspended in buffer (Buffer solutions A and B, see Curds and Cockburn, 1968). The washed cultures were then mixed and left overnight at 25°. During this period most of the *T. pyriformis* were eaten and most of the *T. patula* transformed into the macrostome state. The resultant culture of macrostome organisms was then inoculated into the second stage which contained a small volume of *T. pyriformis* cells that had previously been pumped over from the first stage. When the culture was cleared of *T. pyriformis* cells, the pump between the two reactors could be operated so that a constant suspension of prey was supplied to the predator. Using these methods it has been possible to continuously culture *T. patula* for extended periods of time.

Theory and Use of the Coulter Counter

Theory

The theory of the Coulter Counter is well known and documented (Coulter Electronics Inc., 1957; Kubitschek, 1969) and here only those points of direct interest to our application of the apparatus will be discussed. In essence, a suspension of particles (unicellular organisms, etc.) in a conductive fluid is passed through a small orifice that has electrodes suspended on either side. A current is passed between the electrodes and the resistance caused by the orifice is monitored. Passage of a particle through this sensitive zone causes an increase in resistance by the displacement of its own volume of electrolyte from that zone. The change in resistance is related to the volume of the particle and when the equivalent spherical diameter of the particle lies approximately between 2 and 40% of the orifice diameter the relationship is linear.

Coincidence

Several workers have commented upon the statistics of the coincident passage of several particles through the sensing zone and upon the consequent effect on the number of particles actually recorded. Princen and Kwolek (1965) reviewed previous techniques for coincidence correction and derived their own method. However, their derivation relies upon the assumption that particles arriving in the sensing zone will be a random event (i.e. a linear distribution function) which is in conflict with the results of Roach (1968), who stated that the distribution function will be exponential. Using such a function we have derived the expression

$$n = N\,e^{-\psi N} \tag{1}$$

where n is the observed count, N is the actual number of suspended particles and ψ is the coincidence correction constant (which is in the order of 10^{-6}). The derivation of Eqn (1) is given in Appendix 1 on p. 176 and a FORTRAN program for its solution, for reasonable values of n and ψ is given in Appendix 2.

The value of the coincidence constant, ψ, must be estimated for each set of glassware used. This is achieved by setting up a dilution series of particle suspensions, each of which is counted with the Coulter Counter. The dilution factor, d, may be defined by the expression

$$d = \frac{N_d}{N} \tag{2}$$

where N_d is the actual number of particles in the suspension with dilution d and N is the actual number of particles in the stock suspension.

From Eqns (1) and (2) the observed counts at each dilution, n_d, will be

$$n_d = N\,d\,e^{-\psi N d} \tag{3}$$

from which

$$\ln\left(\frac{n_d}{d}\right) = \ln(N) - \psi\,N\,d. \tag{4}$$

Examination of Eqn (4) shows that if our assumptions are correct then a plot of $\ln\left(\frac{n_d}{d}\right)$ against d will be linear, with a slope of $-\psi N$ and will intercept the $\ln\left(\frac{n_d}{d}\right)$ axis at $\ln(N)$. From the values of the slope and intercept it is possible to evaluate the coincidence constant ψ. We achieved a standard deviation of 0·998 when we performed this plot whereas a standard deviation of 0·980 was obtained for the Princen and Kwoleck (1965) method. The method published by Coulter Electronics Ltd (1968), however, gave a standard deviation of only 0·680. It may be concluded therefore that there is no significant difference between our correction and that of Princen and Kwoleck (1965) within the count range studied but both were significantly better than the method suggested by Coulter Electronics Ltd (1968).

Princen and Kwoleck (1965) correctly pointed out that there are two types of coincidence. Type 1 occurs when a particle is shadowed or obscured by the presence of another particle within the sensing zone.

Type 2 occurs when two or more particles, whose size would not individually exceed the threshold size, create a single pulse of additive volume, which does exceed the threshold. Type 1 coincidence results in counts being lowered, while Type 2 results in the counts being raised.

When the Coulter Counter is being used purely as a counting device, the threshold is set at a level below the size of the smallest of the particles to be counted. In this case Type 1 coincidence should be corrected by the method outlined above (also see Appendix 2 on p. 176) and Type 2 coincidence will be negligible since there are few sub-threshold size particles. However, if volume distributions are to be derived using the Coulter Counter, then Type 2 coincidence may cause major inaccuracies. The statistics of Type 2 coincidence are much more difficult and we have pursued the mathematics no further since the use of a Coulter Channelyzer overcomes the problem. Smither (1975) has used a Coulter Counter Model Fn to derive growth curves and volume distributions for *Escherichia coli* with a rapid and simple technique (ignoring Type 2 coincidence effects) which agree well with optical densities and plate counts.

The Channelyzer receives pulses from the Coulter Counter and assigns each pulse to one of a hundred channels according to its size, in this way a complete volume distribution is automatically generated. As the coincident passage of particles produces a mis-shapen pulse, these can be edited out electronically and therefore all coincidence can be ignored in the computation of the volume distribution using the Channelyzer. The mean cell volume (MCV) of the particles in the distribution can be calculated from

$$\text{MCV} = \frac{\sum n_c C}{\sum n_c} \tag{5}$$

where n_c is the number of particles in each channel and C is the volume represented by each channel.

Volume measurements

There has been much discussion in the literature concerning the relationship of particle shape to the observed volume as determined by the Coulter Counter (Anderson *et al.*, 1967; Batch, 1964; Bloch and Gusack, 1963; Eckhoff, 1967; Gregg and Steidley, 1965; Grover *et al.*, 1969; Harvey and Marr, 1966; Hurley, 1970). Most authors agree that the effect of size gives rise to less than 8% error in the true volume. However, in the case of *Tetrahymena pyriformis*, while estimations of the volume of this organism by optical methods (microscopically measuring length, breadth and assuming the shape to be a prolate spheroid) lie in the range 8000–50 000 μm^3 (Curds and Cockburn, 1971), estimates obtained using a Channelyzer are in the range 1000–11 000 μm^3. The great difference between the two volume estimates is probably due to a combination of two factors, firstly the shape of *T. pyriformis* is not truly that of a prolate spheroid but tends to be flattened, and secondly in our experience *T. pyriformis* feeding on bacteria are larger than those grown axenically. The MCV of an axenic population of this ciliate has now been measured both optically and electronically over a wide size range and it was found that the former method is on average 3·3 times greater than the volume measured by the latter method.

On repeated, rapid, independent estimations of MCV using the Coulter Counter we have achieved a reproducibility of better than 1% (based on six observations). There is a linear relationship between the volume of a cell measured by the Coulter Counter, its volume as measured by optical methods and the dry weight of the cell, therefore we have concluded that the Coulter Counter technique is the most accurate method available to estimate the biomass of ciliates particularly when more than one species is present in the population.

Effect of diluent upon MCV

Morrison and Tomkins (1973) found that the dilution of cells prior to counting or sizing had a profound effect on their MCV. The volume of *Tetrahymena* rapidly increases (within 3 min) and decreases back to within 1% of their original volume following dilution in 6·5 g $litre^{-1}$ sodium chloride solution. From our own experience both volume and population numbers are then maintained constant for a further 40 min before permanent swelling occurs. We have tried many fixatives to preserve the shape and size of *Tetrahymena* without success and therefore count and estimate MCV of our cells while alive. Since the cells are motile this method has the added advantage of preventing sedimentation which can cause inaccurate sampling.

Volume distributions in Tetrahymena

A typical volume distribution of *T. pyriformis* obtained with a Coulter Counter Model Fn and Channelyzer is shown in Fig. 2. In an actively growing culture of micro-organisms, each old (large) cell divides to produce two young (small) cells. Therefore the result will be the presence of more small cells in the culture than large cells. This is demonstrated by the shape of the distribution in Fig. 2 which is skewed with a long

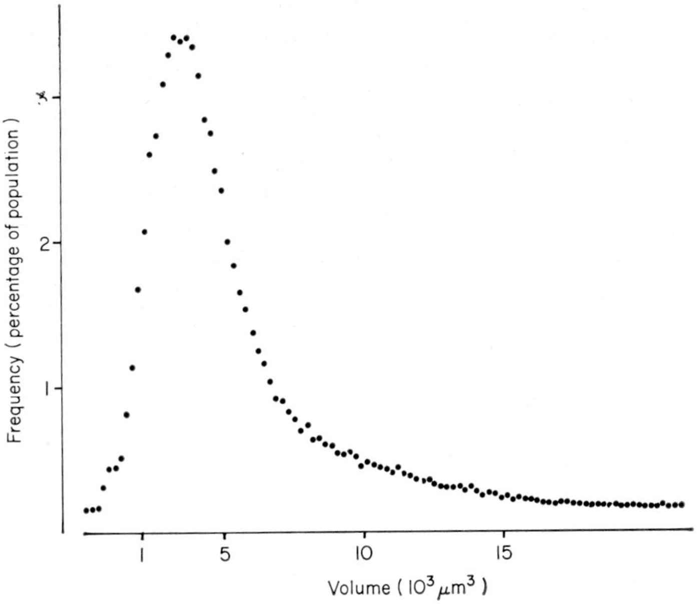

FIG. 2. Volume distribution of *Tetrahymena pyriformis* taken from an exponentially growing culture.

tail of larger cells. The Γ-distribution may be of value in an empirical fit to this data and a common form of this distribution (Hastings and Peacock, 1974) is given by the expression

$$f(x) = \frac{\left(\frac{x}{b}\right)^{c-1} e^{-\frac{x}{b}}}{b\,\Gamma(c)} \tag{6}$$

where x is the size of the cell, b is the scale factor and c the shape factor. However, since growing cells cannot achieve zero size the Γ-function is displaced from the origin. For this reason a further constant, the displacement factor, z, must be incorporated into Eqn (6), so that

$$f(x) = \frac{\left(\frac{x-z}{b}\right)^{c-1} e^{\frac{z-x}{b}}}{b\,\Gamma(c)}. \tag{7}$$

We are currently attempting to develop a digital technique to fit experimental data to this expression using the mode, mean and some other property of the curve for the estimation of the shape, scale and displacement factors.

The bimodal curve shown in Fig. 3 is caused by the dual morphological form of *T. patula* and is a combination of the distributions of the microstomes and macrostomes, of which each population would individually have the same form as that in Fig. 2. While methods exist for the separation of normal distributions, to our knowledge, none exist for the separation of overlapping Γ-distributions. The technique of moments (Pearson, 1895) is unreliable for heavily skewed distributions, as each successive moment gives increasing bias to the tail of the distribution. We hope to be able to use a maximum likelihood method for the eight parameters that need to be evaluated for two overlapping Γ-distributions.

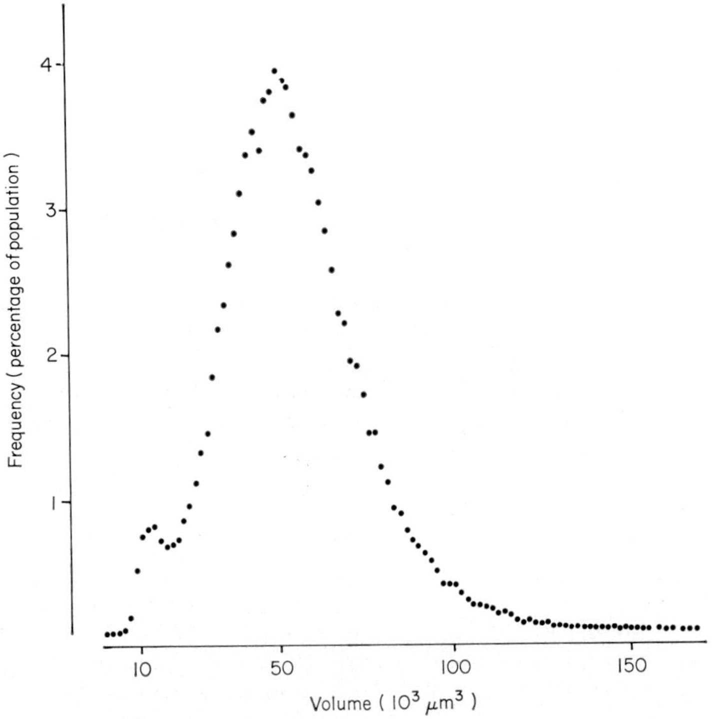

FIG. 3. Volume distribution of *Tetrahymena patula* taken from late-log phase of an axenic batch culture. The small peak represents the microstomes and the large peak the macrostomes which predominate at this stage of a culture.

Acknowledgement

We wish to record our thanks to Dr. Michael Hills of the Biometrics Section, British Museum (Natural History) without whose aid the development of the mathematical procedures mentioned in this paper could not have been attempted.

References

ANDERSON, E. C., PETERSON, D. F. & TOBEY, R. A. (1967). An effect of cell shape on apparent volume as determined with a Coulter aperture. *Biophys. J.*, **7**, 975.

BATCH, B. A. (1965). The application of an electronic particle counter to size analysis of pulverized coal and fly-ash. *J. Inst. Fuel*, **37**, 455.

BLOCH, M. G. & GUSACK, J. A. (1963). Denier distribution of textile fibres obtained with an electrical sensing zone particle analyzer (Coulter Counter). *Text. Res. J.*, **33**, 224.

COULTER ELECTRONICS INC. (1957). Theory of the Coulter Counter. Bulletin T–1, Appendix 1. In *Instructions and service manual for the Coulter Counter.*

COULTER ELECTRONICS LTD (1968). *Instruction and service manual for the Model "Fn" Coulter Counter* Section III, Pt 3–2·2 (supplement, p. 8).

CURDS, C. R. & BAZIN, M. J. (1977). Protozoan predation in batch and continuous culture. In *Advances in aquatic microbiology*, Vol. 1 (Droop, M. R. & Jannasch, H. W., eds). London and New York: Academic Press, p. 115.

CURDS, C. R. & COCKBURN, A. (1968). Studies on the growth and feeding of *Tetrahymena pyriformis* in axenic and monoxenic culture. *J. gen. Microbiol.*, **54**, 343.

CURDS, C. R. & COCKBURN, A. (1971). Continuous monoxenic culture of *Tetrahymena pyriformis. J. gen. Microbiol.*, **66**, 95.

ECKHOFF, R. K. (1967). Experimental indication of the volume proportional response of the Coulter Counter to irregularly shaped particles. *J. scient. Instrum.*, **44**, 648.

ELLIOTT, A. M. (1973). *Biology of* Tetrahymena. Stroudsburg, Pennsylvania: Dowden, Hutchinson and Ross, Inc.

GREGG, E. C. & STEIDLEY, K. D. (1965). Electrical counting and sizing of mammalian cells in suspension. *Biophys. J.*, **5**, 393.

GROVER, N. B., NAAMAN, J., BEN-SASSON, S., DOLJANSKI, F. & NADAV, E. (1969). Electrical sizing of particles in suspensions. II. Experiments with rigid spheres. *Biophys. J.*, **9**, 1415.

HAMILTON, R. D. & PRESLAN, J. E. (1970). Observations on the continuous culture of a planctonic phagotrophic protozoan. *J. exp. mar. Biol. Ecol.*, **5**, 94.

HARVEY, R. J. & MARR, A. G. (1966). Measurement of size distributions of bacterial cells. *J. Bact.*, **92**, 805.

HASTINGS, N. A. J. & PEACOCK, J. B. (1974). *Statistical distributions.* London: Butterworth.

HILL, D. L. (1972). *The biochemistry and physiology of* Tetrahymena. New York and London: Academic Press.

HURLEY, J. (1970). Sizing particles with a Coulter Counter. *Biophys. J.*, **10**, 74.

JOST, J. L., DRAKE, J. F., FREDRICKSON, A. G. & TSUCHIYA, H. M. (1973). Interactions of *Tetrahymena pyriformis*, *Escherichia coli*, *Azotobacter vinelandii* and glucose in a minimal medium. *J. Bact.*, **113**, 834.

KUBITSCHEK, H. E. (1969). Counting and sizing micro-organisms with the Coulter Counter. In *Methods in microbiology*, Vol. 1 (Norris, J. R. & Ribbons, D. W., eds), London and New York: Academic Press, p. 593.

KUBITSCHEK, H. E. (1970). *Introduction to research with continuous cultures.* Englewood Cliffs, N.J.: Prentice-Hall Inc.

LOEFER, J. B., OWEN, R. D. & CHRISTENSEN, E. (1958). Serological types among

thirty-one strains of the ciliated protozoan *Tetrahymena pyriformis*. *J. Protozool.*, **5**, 209.
MALEK, I. & FENCL, Z. (1966). *Theoretical and methodological basis of continuous culture of microorganisms.* New York and London: Academic Press.
MORRISON, G. A. & TOMKINS, A. L. (1973). Determination of mean cell size of *Tetrahymena* in growing cultures. *J. gen. Microbiol.*, **77**, 383.
PEARSON, K. (1895). Contributions to the mathematical theory of evolution. *Phil. Trans. R. Soc.*, Ser. A., **185**, 71.
PIRT, S. J. (1975). *Principles of Microbe and Cell Cultivation.* Oxford: Blackwell.
PRINCEN, L. H. & KWOLEK, W. F. (1965). Coincidence correction for particle size determination with the Coulter Counter. *Rev. scient. Instrum.*, **36**, 646.
PROPER, G. & GARVER, J. C. (1966). Mass culture of the protozoa *Colpoda steinii*. *Biotechnol. Bioengng.*, **7**, 287.
ROACH, S. A. (1968). *The theory of random clumping.* London: Methuen.
SMITHER, R. (1975). Use of a Coulter Counter to detect discrete changes in cell numbers and volume during growth of *Escherichia coli*. *J. appl. Bact.*, **39**, 157.
WILLIAMS, N. E. (1960). The polymorphic life history of *Tetrahymena patula*. *J. Protozool.*, **7**, 10.

Appendix 1

A series of short overlapping dashes of equal length scattered randomly on an infinitely long line can be considered to be a useful analogue of the Coulter Counter coincidence phenomenon (see Chapter 3 in Roach, 1968, for a full discussion). Each dash represents the duration of each particle within the orifice zone and the infinite line symbolizes the time axis.

If each dash is of length l, and if there are N dashes per unit length of line (that is the length of the counting period) then the mean number of dashes in any randomly chosen length l of this line will be Nl. From the Poisson distribution it may be deduced that the probability of there being no dashes in any length l can be given by the expression

$$\exp(-Nl) \tag{A1}$$

For Type 1 coincidence, a particle will be counted by the Coulter Counter only if it enters an empty orifice zone, therefore only the first dash will be counted in any group of overlapping dashes. Thus the probable number of particles counted will be the product of the probability of occurrence of a dash (i.e. N) and the probability that there was no dash within the length l before it, that is,

$$N\exp(-Nl) \tag{A2}$$

Appendix 2

A FORTRAN program for the solution of Type 1 coincidence is given below. Any complete program should include these lines in order, other program lines may be added where spaces are indicated. For example, a simple loop could be constructed to generate a table of coincidence correction. In the following program C and A denote the coincidence constant ψ and the number to be corrected respectively.

```
   •
   •
   •
   DOUBLE PRECISION A, B, C, H, R, T, Q
   •
   •
   •
   READ ( , ) C
 1 •
   •
   •
   READ ( , ) A
   I = 1
   R = C * A
   Q = DEXP (R)
   B = A * Q
 2 H = B
   IF (I .GT. 500) GO TO 3
   I = I + 1
   R = C * H
   Q = DEXP (R)
   B = A * Q
   T = B - H
   IF (T .LT. 0·00) T = -1·0 * T
   IF (T .GT. 0·001) GO TO 2
   WRITE ( , ) B
 3 CONTINUE
   •
   •
   •
   GO TO 1
   •
   •
   •
```

Unit System for Selection of Mixed Interactive Cultures for Industrial Steady-state Fermentations

V. F. LARSEN, R. S. HOLDOM, M. J. SPIVEY AND M. TODD

University of Strathclyde, Glasgow, Scotland

"A culture will give optimal performance only if it has been selected under environmental conditions (pH, temperature, carbon and nitrogen source, etc.) specifically designed for the desired application." (Johnson, 1972)

Introduction

Although industrial uses of microbes always have specific applications in view, Johnson's advice is often neglected during the design of culture screening programmes. When the initial culture enrichment and selection programmes use simple traditional culture systems such as bottles and tubes, or static or shaken flasks containing laboratory culture media, there is a high risk that the required industrial performance cannot be extracted from the derived cultures during pilot or full-scale fermentations (Hockenhull, 1975). Also, the range of species made available for secondary screening of cultures for some desired property is inevitably restricted in traditional microbiological methods which are based on pure culture isolations (Fig. 1); a wide range of microbial interactions is excluded from consideration. This is regrettable since in nature symbiotic and other interactions between microbes are the rule rather than the exception (Bungay, 1968). Applications of mixed cultures in industry are few, although the fermentation of dairy products (Harrison *et al.*, 1975) and the anaerobic digestion of wastes (Ainsworth, 1973) have been particularly successful. More empirical uses of mixed cultures include composting (Gray *et al.*, 1971) and aerobic waste treatments (Harder, 1975; Jannasch and Mateles, 1974; Meers, 1973). In these processes inoculation, culture selection and maintenance are usually left to occur naturally. However, directed uses of mixed cultures by industry are beginning to appear in the literature (see Appendix 1 on p. 211). Many of the new fermentations planned for the next few decades (Davies, 1974) will

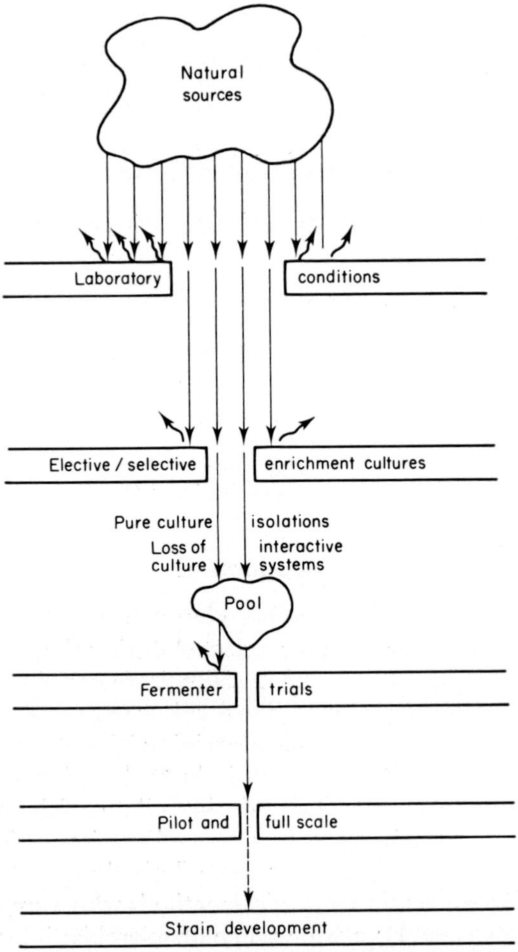

FIG. 1. Schematic representation of traditional methods for isolation and screening of microbial cultures for industrial use. Note that unless the selection conditions reflect the final use conditions, there can be no guarantee that the derived cultures will be suitable for the desired purpose. Also, the potential value of mixed interactive cultures is lost very early.

feature non-aseptic processes in which the performance and stability of the mixed cultures will need to be controlled closely and carefully optimized. Table 1 references some of the advantages claimed for mixed cultures from an industrialist's point of view. If any of these advantages are to be realized fully in practice, the problems of isolation and characterization of interactive cultures, and also the problem of maintaining such cultures in some preferred steady state must be solved; a new methodology is required. In addition, the current trend towards use of sophisticated process control in production scale fermentations (Demain, 1972; Humphrey, 1972) presupposes that cultures worthy of the process will be available. Hockenhull (1975) suggests that in order to ensure "transferability" between pilot and production scale operations, pilot

TABLE 1. Industrially useful features of mixed cultures in fermentation systems

Advantage over pure cultures	Reference
Complex mixed substrates used more efficiently	Bungay (1968)
Process stability improved	Veldkamp (1975), Jost *et al.* (1973), Harrison *et al.* (1975), Wilkinson *et al.* (1972, 1974)
Yield increased	Miller (1966), Harrison *et al.* (1975), Jannasch *et al.* (1974)
Maximum cell density increased	Vary *et al.* (1967), Jannasch *et al.* (1974)
Generation times lowered	Harrison *et al.* (1975), Vary *et al.* (1967), Sheehan *et al.* (1971), Jannasch *et al.* (1974)
Growth at higher temperatures	Jannasch *et al.* (1974), Snedecor *et al.* (1974)
Food value increased	
lower nucleic acid content	Snedecor *et al.* (1974)
higher amino acid and vitamin spectrum	
Incidence of contamination lowered	Harrison *et al.* (1975)
Non-aseptic conditions can be used	Jannasch *et al.* (1974), Johnson (1972)
Foaming problems reduced	Harrison *et al.* (1975)

plant equipment must in future have "all the basic sophistication of the best and most flexible production plant plus the extra refinements that are needed for the creation of new techniques altogether". Our thesis is that the same level of sophistication should be used during the stages of isolation and development of a culture so that transferability is not marred by initial selection of inappropriate cultures.

The unit process described here has been used for the isolation and characterization of mixed cultures on natural gas (methane) but it is offered as a flexible system ready for application to numerous aerobic and anaerobic fermentations, whether based on pure or mixed cultures.

In the context of industrially useful cultures, our fermentation system is designed to satisfy five main objectives

(1) to permit a choice of operating conditions (Table 2) which are accurately controlled during isolation of cultures;

(2) to produce cultures which are amenable to use in scaled-up reactors, thus bypassing in part or whole the normal strain development trials;

(3) to capitalize on the wide genetic potential of mixed interactive cultures by pressurizing the cultures constantly towards the limits of their performance under the chosen environmental conditions (Fig. 2);

(4) to provide an early indication of the best performance likely to be obtained with a given culture. In this way cultures with undesirable features for plant use are quickly recognized and can be discarded;

(5) to provide mass-balance data for the products of fermentation (Johnson, 1972).

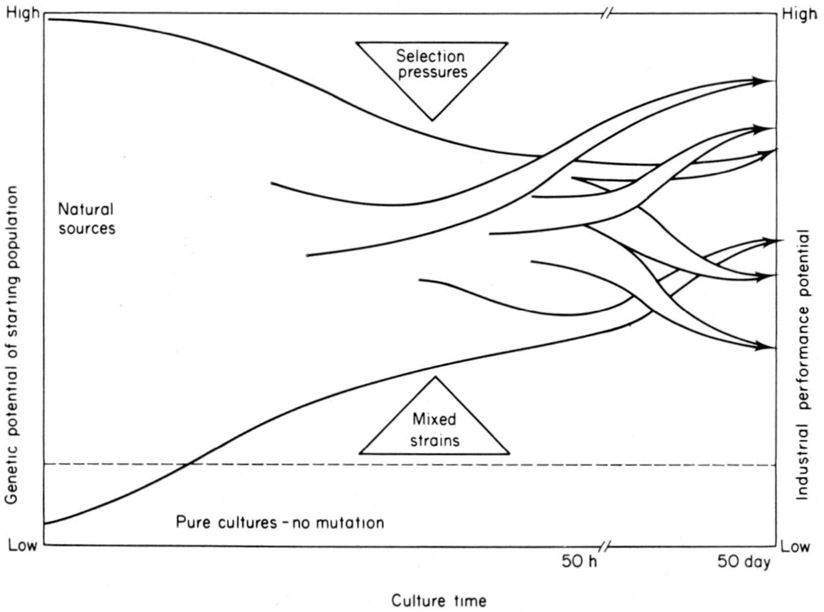

FIG. 2 Continuous culture as an open system for selection of mixed interactive species of industrial value. Continuous-culture techniques are used to simulate the industrial use conditions relevant to a process. The environmental selection pressures bring about a shift in the microbial balance of the starting population to suit, as far as genetically possible, the new ecological niche which now incorporates features of industrial interest. Mixed cultures clearly possess a higher genetic potential at the outset than do pure cultures of natural origin. On average, about 50 days of continuous culture represents full exploration of the genetic potential of the starting population in the imposed conditions. The industrial performance of the derived culture may then be predicted with some confidence and optimized by careful manipulation of the selection pressures.

The purpose of this article is to provide the technical information required to allow successful adoption of the unit system by other workers, and to show how four yield factors of crucial importance in describing the industrial performance profile of a culture can be derived from the continuous record of data from a typical methane fermentation.

TABLE 2. Control options and ranges of instrumentation for gas-recycle continuous methane fermentation

Gas partial pressures	Gas make-up flows*
CH_4 0–100% v/v	CH_4 1–15 litres h^{-1}
O_2 0–100% v/v	O_2 1–30 litres h^{-1}
CO_2 0–30% v/v	CO_2 1–5 litres h^{-1}
Aeration efficiency	
(affects cell density)	Agitation
K_La = 2500 units h^{-1}	250–3000 r/min
Fermenter overpressure	Pressure control
1·4–34·5 kPa (10·2–5 p.s.i.)	10–45 cm water gauge
Gas circulation rate	
60–252 litres h^{-1} (0·2–3·5 v.v.m.)	
Dissolved oxygen tension	
0–133 000 Pa (0–1000 mmHg)	
(approximately 5 times air saturation levels)	
Dilution rate (liquid medium)	Generation times
0·05–2·0 h^{-1}	20 min–14 h
Culture volume	Cell productivity
4·5–5·5 litres ± 1% on control	0–10 g h^{-1}
Limiting nutrient in chemostat mode	
C (as CH_4)	
C (as soluble nutrient, e.g. methanol)	
O (as dissolved O_2)	
N (as NH_4^+ or NO_3^+)	
P (as phosphate)	
Mg (as $MgSO_4$)	
Zn (as $ZNSO_4$)	
Temperature	pH
20–70°	3–12 (limit 6·9 with our medium)

* Limited only by choice of orifice plate size and valve size.

Methods and Materials

Basic features of the unit process

The fermentation unit (Fig. 3) consists of two 7-litre fermenters forming part of a closed gas circulation loop in which oxygen, methane and carbon dioxide are circulated by means of a pump. Each fermenter can be supplied with gas individually, or both fermenters can be supplied either in parallel or in series. Oxygen consumed by the growing culture is balanced by a measured flow of oxygen into the circulation loop. Gas concentrations in the loop are analysed at three sample points: prior to the fermenter train; between each fermenter and downstream of the fermenters. The inlet and outlet gas concentrations of each fermenter can therefore be measured (or calculated) directly for all modes of operation of the fermenters.

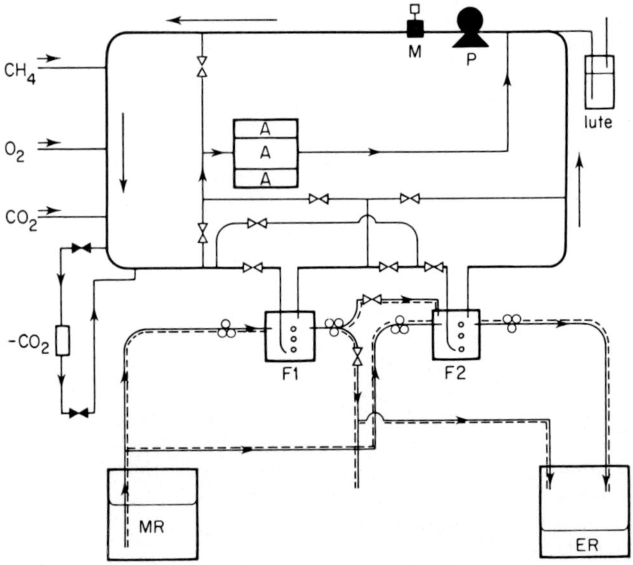

FIG. 3. General arrangement of supply lines for gases and liquids to a two stage multistream, gas-recycle continuous culture. A, analysers for CH_4, O_2, CO_2; P, gas recycle pump; M, manostat pressure controller; F1, F2, fermenter vessel; MR, reservoir of mineral salts medium; ER, receiver for effluent culture; ——, gas flow lines; ══, liquid flow lines; $_o{}^o{}_o$, peristaltic pump

Since the carbon substrate is gaseous, the supply of liquid medium contains only mineral salts. Each of the two fermenters can be supplied individually or both can be supplied, either in parallel or in series, for multistage continuous fermentation. Each fermenter is equipped with individual control of temperature, pH, liquid volume and gas and liquid flow rates (Fig. 3).

Gas recycle

Practical reasons for recycling the fermenter gases relate to

(1) economy in the consumption of gases which would otherwise be wasted in single-pass flow of gas through the fermenters;

(2) the ability to use the closed system as a large "respirometer" for accurate determination of gas consumptions;

(3) the need for high levels of carbon dioxide in the gas phase for optimal growth of micro-organisms (Pirt, 1975), and more specifically for initiation of growth of methane bacteria from small populations (Vary and Johnson, 1967; Dworkin and Foster, 1956; Leadbetter and Foster, 1958).

Safety considerations

A few of the features described here are essential for safe use of methane/oxygen gas mixtures and can be omitted in other applications. However, since fermenter gases are recycled in all cases, pure oxygen must be used to replace oxygen consumed by the culture. If air is used as the supply of oxygen, nitrogen gas builds up in the system and prevents control of the gas composition. Special precautions are necessary when using pure oxygen, especially if the substrates and/or intermediates of fermentation are volatile. Only apparatus which is fit for oxygen service is detailed below. All apparatus and pipework not specifically supplied for oxygen service by the manufacturer must be solvent-degreased with perchloroethylene (Spivey, 1973). All parts of the system involving moving parts (e.g. fermenter vessels) or fluids in motion (e.g. gas pipelines) must be carefully earthed against build-up of static electricity. Special attention is needed where rubber gaskets or similar insulating materials may break the continuity of the earthing.

Full details of a comprehensive fail-safe control facility for methane fermentations are given by Spivey (1973).

Size of fermentation vessel

Although the use of small culture vessels (e.g. 250 ml) minimizes the requirement for medium during continuous fermentation, it is essential that vessels of 2–5 litres working volume be used in this system, for five main reasons

(1) the growth which almost always occurs on the walls of the fermenter must be a small fraction of the total population (Johnson, 1972);

(2) control of dissolved gas concentrations by adjustment of the partial pressures of the component gases is not satisfactory in small culture systems with small overall demands for gases;

(3) sampling from the fermenter can disrupt seriously steady-state measurement and control of fermentation parameters unless the

fermenter volume is large compared with the sample volume (Hockenhull, 1975);

(4) gas leaks from fermenter systems arise far more routinely than is realized, especially in vessels not designed for use at elevated pressures. Undetected leaks can, of course, make nonsense of attempts to derive mass balance and kinetic data from a closed system. It is important to make both the overall working volume of gas and the gas consumption large in proportion to the leak rate in the system. It must be possible to quantify the leak rate at any time during fermentation, especially after making or breaking line joints in the system. Leaks are readily detected if a little freon is circulated under pressure in the system and a halogen detector (Model H 10, Dean and Wood, London) used to "sniff" all joints. The total leak rate is measured directly with the pressure control equipment in the gas recycle system (see below);

(5) industrial grade process control instruments are sometimes the only type available for research scale fermentations and there is a lower limit to their range of measurements; this limit determines the minimum size of fermentation vessel which must be used in a given application.

Medium supply

Use of relatively large vessels for laboratory continuous cultures makes heavy demands on steam-sterilizing facilities in a laboratory and can be

TABLE 3. Composition of industrial salts medium

A	Phosphoric acid (wet process, decanted)	0·0165 M
	Ammonium sulphate (Spanish, 99% purity)	9·0 g
	Magnesium sulphate (Anhydrous, 98% purity)	0·3 g
B	Zinc sulphate ($ZnSO_4.7H_2O$)	0·5 mg
C	As impurities in other ingredients	
	$FeSO_4$	5·00 mg
	$CuSO_4$	0·10 mg
	Boric acid	0·07 mg
	$MnSO_4$	0·50 mg
	Na_2MoO_4	0·10 mg
	$CaCl_2$	13·20 mg
	$CoCl_2$	0·10 mg
	Tap water or deionized water	1 litre
	pH as made	2·6

Chemicals A made at 60 times strength in 500 ml warm water, with stirring; B added after 30 min; mixture diluted to 1 litre with warm water, filtered (Whatman No. 1) and stored until required for dilution to 60 litres in bulk tank.

extremely tedious. An automatic filtration unit for the supply of up to 300 litres day^{-1} of mineral salts medium (Table 3) was developed to overcome these difficulties and forms an integral part of our unit (Larsen *et al.*, 1976).

Use of multistage, multistream continuous culture

The full range of applications of multistage continuous culture systems cannot be covered here; the definitive work by Ricicia (1969) should be consulted. Industrially important uses of multistage systems include the ability to simulate most of the physiological conditions pertaining in all parts of the traditional batch growth cycle, the sequential or simultaneous consumption of mixed substrates in a medium, and the restoration of culture productivity in strains of poor metabolic or genetic stability.

Process control

Successful application of the unit system to a particular strain selection programme requires an understanding of the value of process control in setting and manipulating the environmental conditions which can be imposed on a developing culture. The aim of process control is normally to maintain a process variable such as methane or oxygen concentration at some desired value irrespective of the natural trends for change in the process. However, with just a few extra components, most control systems can also provide valuable data on the rates of change of key process variables in terms of the amount of control agent (e.g. methane and oxygen flows) required to restore the desired value. Such data form a part of the "performance profile" of the culture under development (see below).

Basic control loop

All control loops consist of at least three parts (Fig. 4): the measuring element (e.g. pH electrode; resistance thermometer), a controller (e.g. blind pH meter with control relays; resistance bridge circuit with relays), and a final control element (e.g. alkali pump; immersion heater). The value of the controlled variable may also be indicated on a meter or recorder chart.

The final control element is, in the majority of cases, a valve which may be designed to operate as an on/off valve (i.e. either fully open or fully closed) or as a proportional valve (i.e. open by an amount proportional to the extent by which the controlled variable deviates from a desired value).

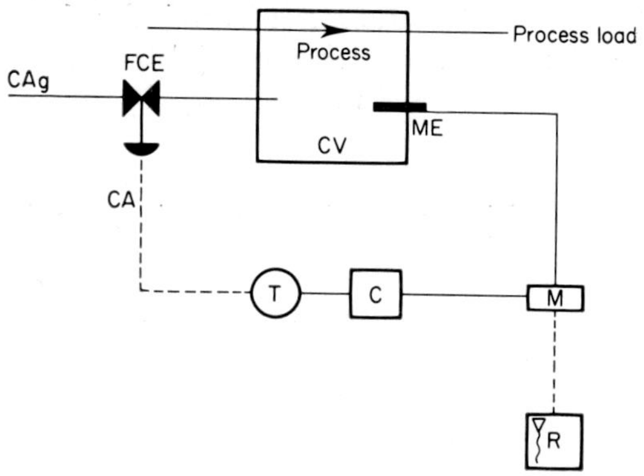

FIG. 4. The features common to all process control loops. The process is represented by a box and includes the fermenter vessel and all ancillary equipment, including gas analysers remote from the fermenter. CV, controlled variable (e.g. pH); ME, measuring element (e.g. pH probe); M, meter for measuring deviation from desired value; R, recorder; C, controller responding to the amount of deviation from the desired value; T, transducer (often integrated with the controller) to generate the desired CA; CA, control action (electrical or pneumatic) delivered to the FCE; FCE, final control element (e.g. valve); CAg, control agent (e.g. alkali). Process load is the instantaneous demand for control agent. A process load change (i.e. a change in the rate of demand for control agent) requires a corresponding change in the amount of control action. Delays in measurement of deviations of CV from the desired value, and/or delays in restoring CV to the desired value by control action, require sophisticated controllers able to measure and compensate for such delay.

Proportional controllers

The controllers required to supply on/off and proportional control action are fundamentally different from one another. Proportional controllers are preferable for close control of dynamically changing process conditions, such as occur during microbial growth. The controller compares the input signal from the measuring element with the preset desired value and adjusts the final control element to minimize the difference between the measured and desired values. The preciseness of control should be displayed on a chart recorder (Figs 5a and 5b). Simple proportional control becomes inadequate if the process load changes. For example, if the rate of production of acidic metabolites changes during proportional control of pH, the control valve admitting alkali opens by the correct amount for each unit of difference between the measured and desired pH values but does not admit enough alkali to restore the pH in the face of increased acid production; this results in a permanent offset or steady-state error in the controlled variable (Fig. 5c). In such cases

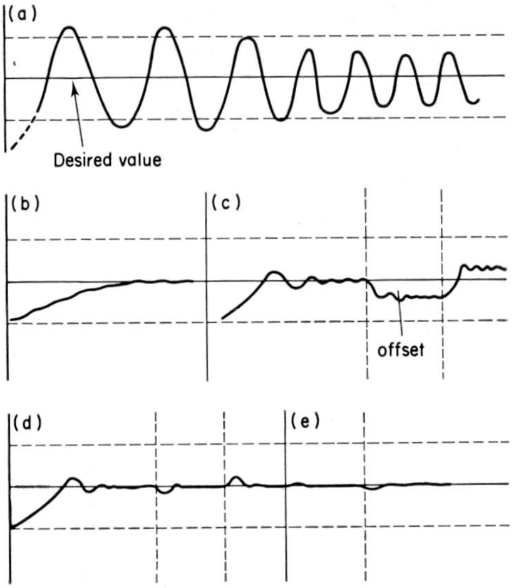

FIG. 5. Three term proportional control of a process subject to changes in process load. a, proportional control (P) only, steady process load, oscillatory control. Band width too narrow; b, bandwidth too wide; c, optimal bandwidth and good control, running into permanent offset caused by large increase and then decrease in the process load; d, P + integral (I) control action removing offset during increase or decrease of process load; e, P + I + derivative (D) control action responding to rate of change of the value of the measured variable. Good control under all conditions.

integral control action is necessary in which the opening of the final control valve is increased by an amount proportional to the period of time for which the variable has remained off-set from the desired value (Fig. 5d). A further refinement is available on most proportional controllers—derivative control action. Here the control action is made larger than normal if the rate of change of the process load, and therefore of the measured value, is high. Similarly, the control action is reduced below normal when the control agent restores the variable towards the desired value at a high rate. The effect is to remove oscillations in the measured value at the expense of a longer time to reach the desired value (Fig. 5e). In continuous cultures at steady state all process loads should by definition be constant and neither integral nor derivative control action is strictly needed. However, all continuous cultures start as batch cultures in which very great changes in process loads occur during growth. In isolation work it is particularly important to maintain precise control of environmental conditions during this early stage of development. With mixed cultures which display oscillations in population density some of the process loads never reach steady-state levels. In such cases three-term proportional control (i.e. including integral and derivative action) is desirable.

The application of these principles to our unit system is now made clear by describing the construction and operation of each control sub-system.

Control of pressure (and, indirectly, methane concentration)

Gas pressure in the system is controlled at three points

(1) the gas leaving the recycle pump at high pressure (Fig. 6) is controlled at 69 kPa (10 p.s.i.g.) over a range of gas delivery rates to the fermenters by a "manostat" regulator (manually adjusted). This provides the constant pressure necessary for satisfactory operation of the flow controllers and flow meters on the gas make-up lines (see below);

(2) the system pressure is then reduced via needle valves (associated with ball-float flow meters serving the fermenters) just enough to maintain the required gas flow through each fermenter. The lowest pressure in the system occurs downstream of the fermenters; a positive pressure at this point ensures that no air gains entry anywhere in the system through leak holes;

(3) the low pressure value (20 cm w.g.) is maintained by three-term proportional control of the flow of one of the fermentation gases (here, methane) and is recorded on a pneumatic recorder. This control loop is crucial to proper functioning of all the other control loops and therefore of the system as a whole. Each gas analyser, for example, produces an output signal which is dependent not only on gas concentration but on gas pressure within the instrument. Variations in gas concentration can only be measured at constant pressure. In all applications of the system not requiring methane, the pressure control loop must be retained; nitrogen should be substituted for methane. If nitrogen is not fixed by the culture the rate of nitrogen make-up will be small and provides a direct measure of the leak rate in the system.

Control of oxygen concentration

The recycled gas is depleted of oxygen after meeting the oxygen demand of the culture. A proportion of the gas supplying the fermenters is continuously bypassed through three gas analysers (Fig. 3), one of which is a paramagnetic oxygen analyser. This is the only type of oxygen analyser which is insensitive to the quantity and type of other gases in

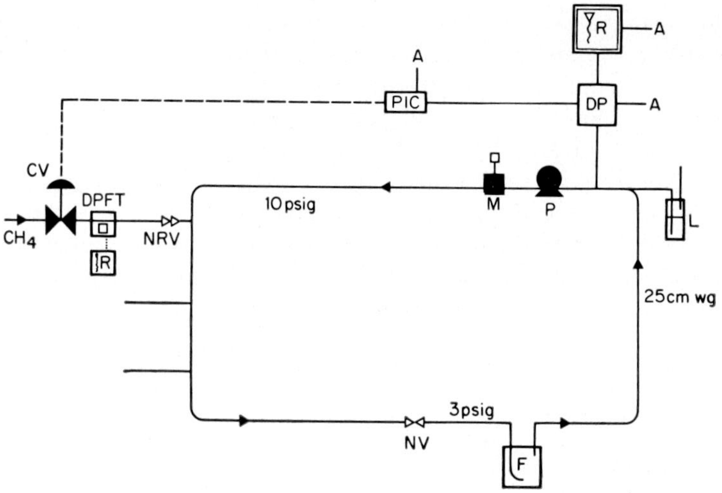

FIG. 6. Control of system pressure

Item	*Description and remarks*	*Supplier*
DP	Model 303 TD001 21 Pressure transmitter 0–51 cm w.g. range	Taylor Instruments, Stevenage, Herts.
PIC	Model 86KF225/402 RF 1041 0–38 cm w.g. linear pressure receiver and controller	Taylor Instruments, Stevenage, Herts.
R	Model 90JF217 Three-pen recorder	Taylor Instruments, Stevenage, Herts.
DPFT	Model 10B3465S Flow transmitter	Fischer and Porter Ltd, Workington, Cumberland
CV	Minim cm air operated valve for 15 litres h^{-1} maximum gas flow, 138 kPa (20 p.s.i.g.) upstream, 34·5 kPa (5 p.s.i.g.) pressure drop. P4 orifice size	G. A. Platon, Basingstoke, Hants
M	Model 66NA Stainless steel fluon-filled pressure regulator, non-bleed type. 800 $litres^{-1}$, 83 kPa (12 p.s.i.g.) upstream, 69 kPa (10 p.s.i.g.) downstream	Electro-Chemical Eng. Co., Sheerwater, Woking, Surrey
P	Model 416–ZE Diaphragm pump	Dawson, MacDonald and Dawson, Ashbourne, Derbs.
NV	Soft seat, $\frac{3}{8}$ in connection needle valve	Don Brown (Brownall) Ltd, Royton, Aldham, Lancs.
A	Oil-free air supply, 138 kPa (20 p.s.i.g.)	
NRV	Non-return valve	
L	Lute	
F	Fermenter	

the gas mixture and is readily calibrated with N_2 (for 0% 0_2) and a suitable span gas (e.g. air for 21% 0_2 on the scale). The analyser output (0–10 mV full scale) serves an electronic three-term proportional controller whose control output (0–100 mV), is converted to a 20–110 kPa (3–15 p.s.i.g.) pneumatic signal suitable for operation of a pneumatic proportional control valve on the oxygen make-up line (Fig. 7). The oxygen concentration is recorded continuously and the proportional controller is finely tuned during a fermentation using the integral and derivative control facilities so that oscillations in oxygen concentration, mostly caused by the inevitable lag between oxygen analysis and oxygen make-up into the system, are minimized.

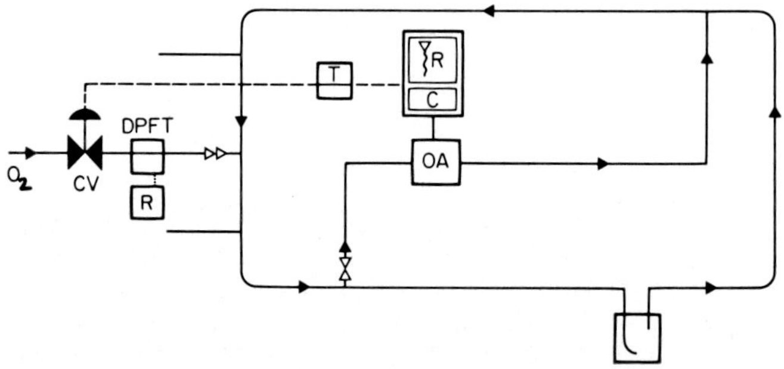

FIG. 7. Control of oxygen concentration

Item	*Description and remarks*	*Supplier*
OA	Model OA137 Paramagnetic oxygen analyser 0–100% 0_2 range, 0–100 mV output	Taylor–Servomex, Crowborough, Sussex
RC	Model 300–930–202–0044–5–001–452 Recorder with three-term proportional current control output (0·5 mA)	Leeds and Northrup, Tyseley, Birmingham
T	Model 701TF111 Transducer 0–5 m A input, 21–104 kPa (3–15 p.s.i.g.) output	Taylor Instruments, Stevenage, Herts.
CV	Minim $\frac{1}{4}$ in air operated valve for 30 litres h^{-1} maximum gas flow, 138 kPa (20 p.s.i.g.) upstream, 34·5 kPa (5 p.s.i.g.) pressure drop. Pl orifice size. Fluon packed for oxygen service	G. A. Platon, Basingstoke, Hants.
DPFT	Differential pressure flow transmitter	
R	Three-pen recorder	

Control of carbon dioxide concentration

In an established fermentation, large quantities of carbon dioxide build up in the gas recycle line. The resultant rise in pressure closes the gas valve linked to the pressure controller and, in our case, starves the culture of methane. The concentration of carbon dioxide should be controlled at some desired value. The ratios of 0_2:CH_4:CO_2 must be carefully optimized if high cell yields are to be realized (Whittenbury, 1969). Our system provides for manual addition of CO_2 to a preset concentration during start up, with further small additions (in practice 100–200 ml day^{-1}) to compensate for losses through system leaks. During continuous fermentations, some CO_2 is also removed from the gas phase by the fresh medium added to the fermenter, and some by the alkali added during control of pH. When the cell productivity exceeds 0·02 g litres^{-1} h^{-1} there is a net build up of CO_2 at a rate depending on the leak rate in the system and control action is initiated as follows. The CO_2 concentration in the gas phase is monitored continuously by an infrared analyser and relayed to a recorder in which adjustable microswitches mounted on the shaft on the slide-wire provide on/off control action about the desired CO_2 value. A multipoint recorder is used (for economy) and "hold-in"

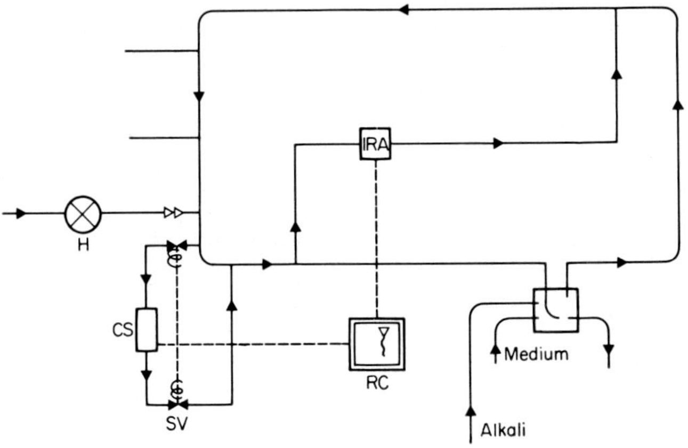

FIG. 8. Control of carbon dioxide concentration

Item	*Description and remarks*	*Supplier*
IRA	Model Irga 20 infra-red analyser 0–30% CO_2 range, 0–100 mV output	Grubb-Parsons, Newcastle-on-Tyne
RC	Model 312–920–504–9189–7–002–304–351 Speedomax H multipoint recorder. Specify circuit board for control of lock-in relays (230 V AC 5 A switching), and two micro-switches for high and low alarm sensing over 0·5–100% of full scale deflection	Leeds and Northrup, Tyseley, Birmingham
SV	Miniature solenoid valves (red-cap) 230 V AC continuous rating, stainless steel	Dewrance-Asco, Skelmersdale, Lancs.
CS	Bypass tube containing soda lime	British Drug Houses, Poole, Dorset
H	Hand valve	

Note: alkali addition (pH control) and fresh medium both remove some CO_2 from the gas phase.

relays are necessary so that control action is held for the CO_2 point while other inputs to the recorder are scanned. The final control elements consist of two solenoid valves on a bypass line from the main gas circulation route. The control agent is non deliquescent soda-lime (B.D.H., Poole, Dorset) contained in rechargeable glass columns in the bypass line (Fig. 8).

Control of methane concentration

The methane concentration in the gas phase is controlled indirectly. The circulating gas is a tertiary mixture of methane, oxygen and carbon

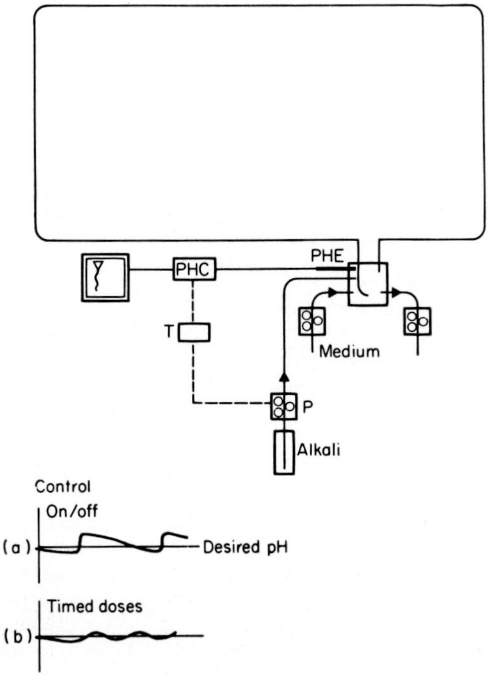

FIG. 9. Control of pH

Item	*Description and remarks*	*Supplier*
PHE	Combination glass/reference electrode with side-arm for connection to 69 kPa (10 p.s.i.g.) air	Activion Glass, Kinglassie, Fife
PHC	Model EIL 91B Meter/Controller	EIL Ltd, Chertsey, Surrey
T	Model 0–8 s variable on- and off-time relay unit	Electroplan, Royston, Herts.
P	Model MHRK fixed speed peristaltic pump	Watson-Marlow, Marlow, Bucks.

dioxide. Oxygen and carbon dioxide are under direct concentration control. Therefore, as long as the pressure of the gas mixture in the system is controlled, the methane concentration is controlled automatically. The methane concentration is monitored by an infra-red analyser.

pH control

The pH of the medium (Table 3) is approximately 3·0 when freshly made up, which is well below the pH range for growth of most bacteria, and is adjusted within the fermenter as part of the pH control function. The low initial pH is a great advantage in preventing growth of contaminants during bulk storage of unsterile medium. For a given dilution rate across the fermenters, a fixed rate of addition of alkali is called for, quite apart from the alkali demand created as a result of growth and metabolism in the fermenters.

Because of the difficulty in obtaining proportional valves for controlling the flow of small volumes (0–10 ml h^{-1}) of strong alkali, on/off control action must be used, preferably with timed interruptions of the flow of alkali from the delivery pump (Figs 9a, b).

Control of dilution rate, fermenter volume and foaming

The growth rate of micro-organisms during continuous cultivation is numerically equal to the residence time of liquid in the fermenter which in turn depends on the liquid volume in the fermenter and the liquid flowrate into the fermenter. Both must be controlled for maintenance of steady-state cultures (Fig. 10).

Liquid flow to the fermenter is controlled by adjustment of the speed of a peristaltic pump. The flow rates are monitored periodically by measuring the volume of medium pumped from the fermenter over a given period.

Control of the liquid volume in the fermenter cannot be achieved by conventional weir overflow devices if industrial levels of gas transfer rate are chosen because the fermenters are completely filled with gas-liquid foam containing microbes; there is no free gas head-space. The gaseous component of the foam must be separated continuously from the liquid before leaving the fermenter by means of a mechanical foam breaker (Chemap, Zurich). However, this in itself does not control the fermenter volume; some measure of the mass of liquid within the fermenter is required. Measurement of the pressure at the bottom of a fermenter would normally reflect the weight of the liquid column therein but measurements are impossible because the mechanical foam breaker

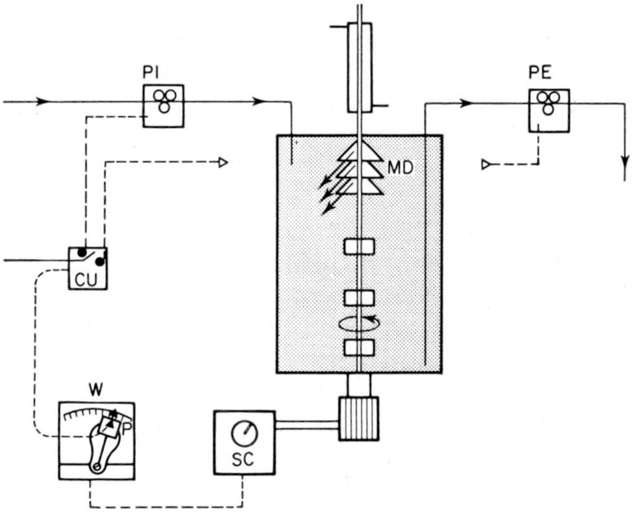

FIG. 10. Control of liquid volume in the fermenter

Item	*Description and remarks*	*Supplier*
SC	Stepless DC speed controller	Chemap., Zurich
W	Wattmeter	Chemap., Zurich
MD	Mechanical defoamer mounted on stirrer shaft	Chemap., Zurich
P	Photocell head }	Chauvin-Arnoux,
CU	Control unit and relay }	Paris
PI	Model MHRE/22 peristaltic pump, variable speed	Watson-Marlow, Marlow, Bucks.
PE	Model MHRE/88 peristaltic pump variable speed	Watson-Marlow, Marlow, Bucks.

creates a variable back pressure in the fermenter. The back pressure in our fermenter—17 kPa (2·5 p.s.i.g.)—is high compared to the pressure exerted by the liquid column in the fermenter, thus making the method too insensitive for control of liquid volume. The foam breaker is driven by the stirrer motor and the power drawn by this motor increases sharply for small increases in the working volume of liquid in the fermenter. This increase in power is measured on an expanded scale wattmeter (0–500 W) which is used to control liquid volume in the fermenter as follows. A photocell head is fitted to the wattmeter dial and adjusted to respond to the change in power. The photocell operates a relay which controls the electrical supply to two peristaltic pumps, one supplying fresh medium the other removing spent medium. One or other of these pumps is always activated, but never the two together. The fermenter volume can be controlled to $\pm$ 1% providing cell density, viscosity, gas hold-up and stirrer speed do not change substantially. Factors affecting the properties of the foam (e.g. temperature, viscosity, surface tension and density) greatly influence the control accuracy. Changes in cell density are unimportant unless very high concentrations of cells are used (e.g. 25 g $litre^{-1}$).

An efficient condenser on the gas outlet line is essential in order to condense water vapour and return the liquid to the fermenter, and also to prevent wetting of outlet filters, gas analysers and other control apparatus in contact with the circulating gas.

Temperature control

The growth rate of micro-organisms in continuous chemostat culture is, within limits, not dependent on temperature. However, the properties of the gas-liquid foam in the fermenter are dependent on temperature, and liquid volume control depends on the properties of the foam. Liquid volume affects dilution rate and therefore growth rate. Solubilities of gases are also temperature dependent. The gas transfer rate, which directly influences the cell density in the fermenter, is in turn dependent

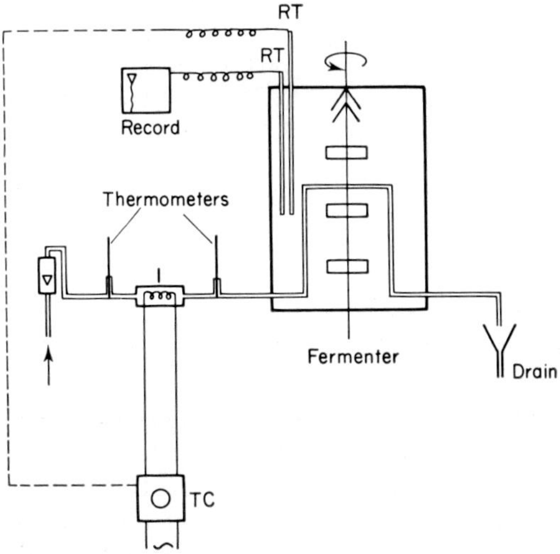

FIG. 11. Control of fermentation temperature

Item	*Description and remarks*	*Supplier*
RT	Resistance thermometer	Eurotherm, Worthing, Sussex
TC	Thyristor controller Model 020 088 70 015 13 00	Eurotherm, Worthing, Sussex
I	Immersion heater; flame-proof model FPCS–25. 2·5 kW, 250 V AC	Eltron (London) Ltd, Croydon, Surrey

on the gas solubilities. It is therefore very important in this system to provide very precise control of temperature. By using a proportional controller, oscillations of temperature about the desired value can be eliminated. The controller varies the electrical power delivered to an immersion heater which is placed in a fixed flow of water to the heat exchanger in the fermenter (Fig. 11). An additional advantage of this control method relates to the ease with which the net heat of fermentation of a culture can be calculated (see below).

Culture performance profile

For industrial production of single cell protein, four yield coefficients are of importance: the yield on carbon source, the yield on oxygen, the heat of formation yield and the carbon dioxide yield. The first three yield factors influence the economics of industrial scale operations; the last yield coefficient is important when the gas is recycled since it affects the control of gas concentrations. The most important yield coefficient is the yield on carbon. This determines not only the amount of carbon source required for a given amount of biomass but also directly affects the yield on oxygen and the carbon dioxide production yield. During single cell protein production the carbon source is converted into biomass, carbon dioxide and water. The larger the amount of carbon which can be incorporated into cell mass the lower the CO_2 product for a given amount of cell mass. Furthermore, since the percentage oxygen content of carbon dioxide is higher than that of cell material, lower carbon dioxide production reduces the overall oxygen demand and increases the calculated yield on oxygen.

The oxygen yield determines the minimum gas transfer rate which must be supplied in the full scale plant and affects the choice of fermenter and its power input.

The heat production yield of the culture determines the requirement for cooling water to maintain a constant fermentation temperature. This yield factor is particularly important where the heat of formation of substrate is large compared with the heat of formation of cells, as it is with methane, because cooling costs then become a major factor in the economic viability of a process.

Alternative performance profiles

Many other factors are normally accounted for in the performance profile of a culture. The relative importance of each factor depends on the ultimate use or application of the culture and is discussed on p. 208.

Mass Balance and Evaluation of Yields

Mass balance for oxygen

The consumption of oxygen by micro-organisms is obtained by taking an oxygen mass balance over the gas circulation system (Fig. 12)

$$\begin{matrix}\text{accumulation of gas} \\ \text{in system}\end{matrix} = \begin{matrix}\text{(input of gas} \\ \text{to system)}\end{matrix} - \begin{matrix}\text{(losses of gas} \\ \text{from system).}\end{matrix}$$

Accumulation is used as a general term covering build up (+) or depletion (—) of gas in the system.

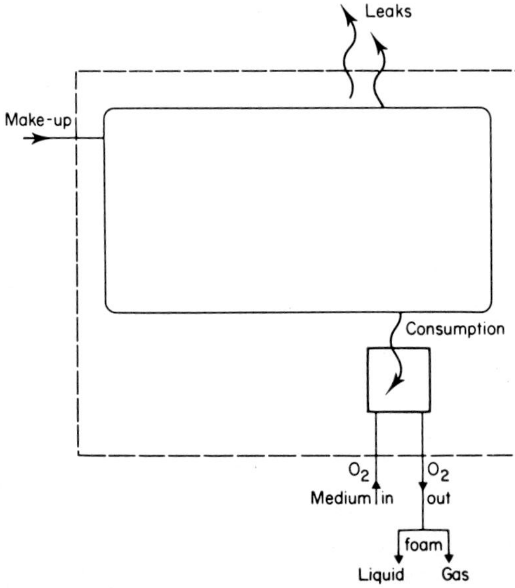

FIG. 12. Mass balance for oxygen.

Accumulation = (make-up flow) + (dissolved oxygen in medium to fermenter) — (dissolved oxygen in medium from fermenter) — (O_2 gas trapped in foam) — (O_2 leaked from system) — (O_2 consumption).

Oxygen is under direct concentration control and there will be no accumulation of gas as long as the system gas volume is constant. Furthermore, the amount of oxygen dissolved in medium passing into and from the fermenter, and the leak rate from the system are extremely small and can be neglected.

The mass balance therefore simplifies to

make-up flow = (consumption) + (loss of gas trapped in foam).

The make-up gas flow is measured as the pressure drop over a 2·5 cm capillary tube during flow of gas through the tube. A differential pressure transducer measures and converts the pressure drop to a pneumatic signal in the 20–110 kPa (3–15 p.s.i.g.) range which is passed to a 3 pen pneumatic recording station.

The loss of gas mixture trapped in the effluent foam can be measured by water displacement in a graduated collection vessel in which the foam separates into gas and liquid phases. Knowing the concentration of oxygen in the gas mixture leaving the fermenter, the net loss of oxygen in the effluent foam can be computed.

Mass balance for methane

A system identical to that for oxygen can be drawn for methane and the same assumption regarding leak rate and dissolved gas level can be made.

However, as the methane is not under direct concentration control accumulation (+ or —) may occur depending on the closeness of control of the concentrations of oxygen and carbon dioxide. If the CO_2 control system is bypassed for $\frac{1}{2}$–1 h during a fermentation the mass balance for methane simplifies to

(accumulation) = (make-up flow) — (consumption) — (loss of methane trapped in foam).

Accumulation of methane in the system is calculated from the observed change in methane concentration over the given period of time. Make-up flow is derived as an average from the relevant section of the recorder chart for methane flow. Loss of methane in the foam is computed as for oxygen. This temporary upset to the methane concentration in the system is only serious when dissolved methane is chosen as the growth-limiting nutrient in chemostat cultures. Oxygen is the preferred limiting nutrient since its concentration in the circulating gas is on analyser control and is not affected by the build-up of carbon dioxide.

Heat Yield

The fermentation of methane to biomass yields large quantities of heat of fermentation. Furthermore, at high stirrer speeds the heat of agitation becomes significant. Both these heat inputs tend to increase temperature whereas heat losses to atmosphere and losses through exit gas and liquid streams tend to decrease the temperature. In our case there is a large net production of heat. The heat of fermentation is directly proportional to

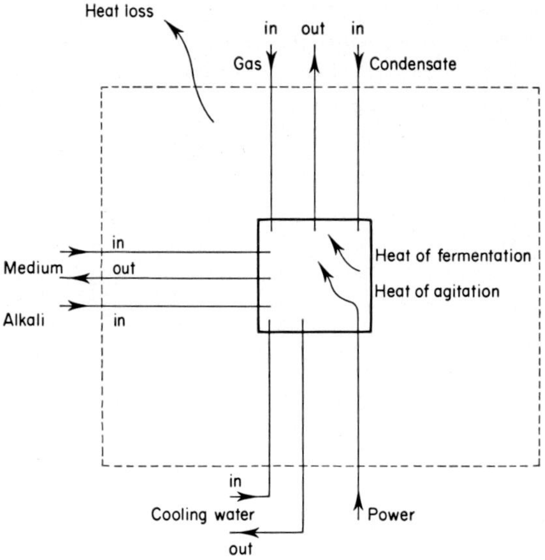

FIG. 13. Heat balance across the fermenter. Determination of the heat of fermentation simplifies measurement of input water flow, and input and output water temperatures, before and during growth of the culture (see text).

the oxygen consumption, and can be obtained during steady-state conditions by determination of the heat balance over the fermenter (Fig. 13).

Heat balance

Since the temperature of the fermenter is controlled there is no heat accumulation. Therefore

heat content of gas input stream	(A)
+ heat content of medium input stream	(B)
+ heat content of alkali input stream	(C)
+ heat content of cooling water input stream	(D)
+ heat of agitation	(E)
+ heat content of condensate return stream	(F)
+ heat of fermentation	(G)

heat content gas outlet stream	(H)
+ heat content of medium outlet stream	(I)
+ heat content of cooling water outlet stream	(J)
+ heat losses to ambient	(K).

The net effect of all heat inputs and outputs except heat of fermentation can be measured by a simulated run prior to inoculation or during the lag period before growth occurs if this is sufficiently long. By rearranging the heat balance equation

$$(J - D) = (A + B + C + E + F) - (H + I + K).$$

The parameters in the right-hand side of this equation are only affected by operating conditions, all of which are held constant in steady-state fermentations. Hence

$$\text{heat of fermentation (G)} = \begin{matrix}(J - D) \text{ during} \\ \text{fermentation}\end{matrix} - \begin{matrix}(J - D) \text{ before} \\ \text{fermentation.}\end{matrix}$$

Parameters J and D are derived from measurements of the flow and the input and output temperatures of the cooling water. Use of proportional temperature control of the fermenter via small adjustments to the temperature of the cooling water greatly facilitates measurement of J and D.

Carbon dioxide yield

The carbon dioxide produced by a culture can be found by two methods, both of which are derived from the mass balance equation for CO_2 (Fig. 14). The amounts of CO_2 entering or leaving the system as dissolved gas

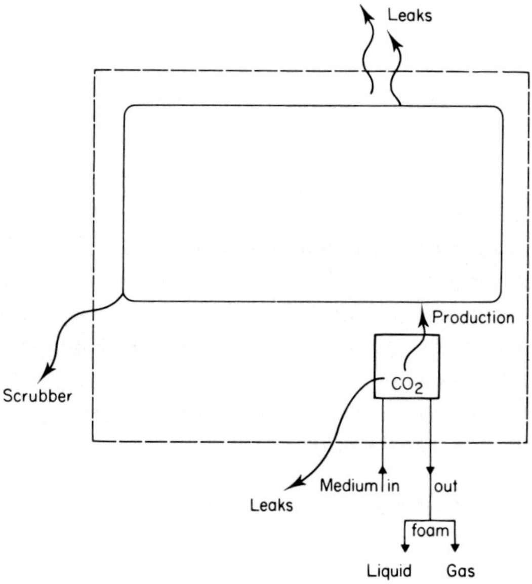

FIG. 14. Mass balance for carbon dioxide.

are negligible with our medium and fermentation conditions. With alternative media and conditions, this point should be checked by measuring decrease in CO_2 concentration in the gas phase caused by delivery of sterile medium at the highest expected dilution rate over a period of time when the pH control system is on, the CO_2 control system is off, and no culture is present in the fermenter.

$$\begin{aligned} \text{Accumulation} &= \text{production} - \text{losses} \\ &= F - (C + G). \end{aligned}$$

In the first method of calculation, the CO_2 control system is assumed to be in use (i.e. there is no accumulation of CO_2). Then

production (F) = loss of CO_2 in foam (C) + CO_2 absorbed in soda lime (G).

C is computed from the total amount of gas in the effluent foam and the average concentration of CO_2 in this gas, as described for oxygen and methane.

G is derived from the increase in weight of the soda lime in the CO_2 scrubber over a given period of steady-state fermentation.

The second method of calculation is particularly useful for instantaneous checks on CO_2 production rate, for example during batch or non steady-state growth of a culture. Parameter G is made zero by temporarily overriding the CO_2 control signal. The rate of CO_2 accumulation is then displayed on the chart recording of CO_2 gas concentration, and is also indicated as a depressed methane make-up flow on the appropriate flow recorder. Then

CO_2 production (F) = accumulation − C.

Results

Performance profile of a mixed culture

Gas containing 75 % CH_4, 10% O_2 and 15% CO_2 was circulated through the fermenter which was held at a constant temperature of 40°.

Five days after inoculation with settled solids from an activated sludge plant, a significant methane consumption occurred. Fresh medium was added at a dilution rate of 0·08 h^{-1} and the culture allowed to reach oxygen limitation, as evidenced by a steady low dissolved oxygen tension reading. The date required for calculation of the performance profile of the culture at this dilution rate are listed in Table 4. Cell productivities over a range of dilution rates are shown in Fig. 15.

TABLE 4. Data obtained during oxygen-limited continuous growth of a methane-oxidizing mixed culture

System gas volume	7 litres
Fermenter liquid volume	5 litres
Average methane concentration	75% v/v
Average oxygen concentration	10% v/v
Average carbon dioxide concentration	15% v/v
Methane make-up flow*	4·2 litres h^{-1}
Oxygen make-up flow	7·1 litres h^{-1}
Rate of change of CH_4 concentration*	–12·5% h^{-1}
Rate of change of O_2 concentration	0% h^{-1}
Cooling water flow rate	0·85 litre min^{-1}
Cooling water inlet temperature†	30·4°
Cooling water outlet temperature†	37·7°
Total gas volume in effluent foam	0·16 litre h^{-1}
Dilution rate	0·08 h^{-1}
Cell density (dry weight)	5·9 g $litre^{-1}$
Weight of CO_2 scrubber at start	323 g
Weight of CO_2 scrubber at finish	346 g
Weight of CO_2 absorbed	23 g
Duration of CO_2 absorption test	5 h

* Determined with CO_2 control system temporarily by-passed. The rate of build-up of carbon dioxide in the system induces a step change in methane make-up flow down to the value listed.

† Equivalent figures on a "blank" fermentation (no culture present) are: 31° (inlet) and 37·6° (outlet).

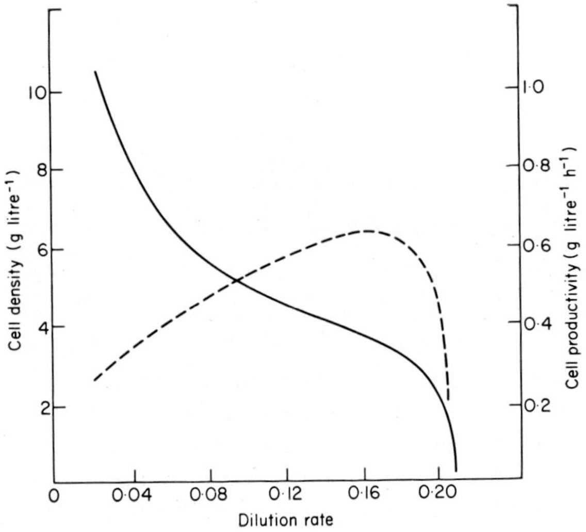

FIG. 15. Cell density and productivity of an oxygen-limited, methane-oxidizing continuous culture over a range of growth rates.

Yield on substrate (Y_{CH4})

Methane accumulation $= \frac{-12 \cdot 5}{100} \times 7$

$= -0 \cdot 87$ litres h^{-1}.

Methane make-up flow $= 4 \cdot 2$ litres h^{-1}.

Methane lost in effluent $= \frac{75}{100} \times 0 \cdot 16$

$= 0 \cdot 12$ litres h^{-1}.

Methane consumption $= 4 \cdot 2 + 0 \cdot 87 - 0 \cdot 12$

$= 4 \cdot 95$ litres h^{-1} at 20°

or $4 \cdot 95 \times \frac{16}{22 \cdot 4} \times \frac{273}{293} = 3 \cdot 33$ g h^{-1} at stp

or, in 1 litre of culture $= 0 \cdot 665$ g $litre^{-1}$ h^{-1}.

Cell productivity $= 0 \cdot 08 \times 5 \cdot 9$.

$= 0 \cdot 472$ g $litre^{-1}$ h^{-1}.

$Y_{CH_4} = \frac{0 \cdot 472}{0 \cdot 665} = 0 \cdot 71$.

Yield on oxygen (YO_2)

Oxygen consumption = oxygen make-up flow

$= 7 \cdot 11$ h^{-1} at 20°

or $7 \cdot 1 \times \frac{32}{22 \cdot 4} \times \frac{273}{293} \times \frac{1}{5} = 1 \cdot 89$ g $litre^{-1}$ h^{-1}.

$YO_1 = \frac{0 \cdot 472}{1 \cdot 89} = 0 \cdot 25$.

CO_2 production yield

CO_2 absorbed = 23 g in 5 h

$= 4 \cdot 6$ g h^{-1}.

CO_2 lost in effluent $= \frac{15}{100} \times 0 \cdot 16$

$= 0 \cdot 024$ litre h^{-1} at 20°

$= 0 \cdot 044$ g h^{-1}.

Total CO_2 produced $= 4 \cdot 6$ g h^{-1}

or $= 0 \cdot 92$ g $litre^{-1}$ h^{-1}.

CO_2 production yield $= 0 \cdot 92$ g CO_2 $litre^{-1}$ h^{-1}

$\div\ 0 \cdot 472$ g cell $litre^{-1}$ h^{-1}

$= 1 \cdot 95$ g CO_2 g cell.

Heat of fermentation

Heat removed by cooling water in absence of culture

$$= 0{\cdot}85\ \text{kg min}^{-1} \times 1\ \text{kcal kg}^{-1}\ \text{deg}^{-1}\ (37{\cdot}6 - 31{\cdot}0)^{\circ}$$

$$= 5{\cdot}6\ \text{kcal min}^{-1}$$

and during fermentation $= 0{\cdot}85 \times 1 \times (37{\cdot}7\text{–}30{\cdot}4)$

$$= 6{\cdot}2\ \text{cal min}^{-1}.$$

Heat of fermentation $= 6{\cdot}2 - 5{\cdot}6$

$$= 0{\cdot}6\ \text{kcal min}^{-1}$$

$$= 36\ \text{kcal h}^{-1}$$

$$= 7{\cdot}2\ \text{kcal litre}^{-1}\ \text{h}^{-1}.$$

$$\text{Heat yield} = \frac{7{\cdot}2}{0{\cdot}472} = 15{\cdot}2\ \text{kcal g}^{-1}.$$

Discussion

There is considerable disagreement over yields of cells on gaseous nutrients such as methane, largely because of the severe experimental difficulties of quantifying gas consumptions and mass balances. This problem greatly hinders theoretical and practical advances in the field (van Dijken and Harder, 1975). In our system gas consumption and mass balance values are obtained with exceptional accuracy since they depend on single measurements of flow made with accurately calibrated instruments operating only under the conditions of calibration. In fermentation systems which do not use gas recycle, these figures must be computed from measurements of gas flow and small differences between inlet and outlet gas concentrations. When gas flows are measured by rotameter devices the compounded error can be as high as $\pm$ 20% (especially when the gas composition and operating pressure vary from those used in the initial calibration of the rotameter) (ICI, Ltd, private communication). Also, the sensitivity of gas analysers is rarely greater than $\pm$ 5% of full range so that the total compounded error can be very large.

An additional advantage of our system concerns the ease with which instantaneous values of gas consumption rate are displayed. The gas make-up flow recorder can be calibrated to read gas consumptions directly as g h^{-1} ($\pm$ 1%). Also, reliable calculation of instantaneous yield factors depends only on accurate measurement of the cell mass in a sample and the liquid medium flow rate to the fermenter. The volume of culture in the fermenter, which varies in most continuous-culture systems and can usually only be determined by temporary interruption of agitation and aeration of the medium and visual estimation of the volume of liquid in the vessel (low accuracy), is not required in our calculation of yields.

The stability of mixed cultures

Mixed cultures will be of greater use to industry if they remain stable over long periods of continuous cultivation.

The Second Law of Thermodynamics predicts that all systems tend towards a state of maximum entropy (greatest disorder) at which point a stable equilibrium exists. The paradox is that life forms constantly struggle against this tendency. Mixed cultures which consist of random cohabiting forms are less highly ordered and less stable than mixed cultures in which one or more interactions, such as symbiosis, have evolved. No culture system, of course, escapes the entropic doom predicted above. However, cultures are able to maintain dynamic and ordered steady states, though only in continuous open systems, by ensuring that the rate of entropy production is minimum for the specific energy flow taking place. In other words their natural remit in an open system is to produce entropy at a minimal rate by maintaining a steady state even in the face of environmental change (the "wisdom of living organisms", Lehninger, 1965).

Most culture isolation programmes unfortunately feature closed systems (e.g. simple batch cultures) or semi-open systems (e.g. serial subcultures) which do not allow, or give insufficient time for this powerful and natural entropic tendency to be realized; nor do they provide growth conditions of sufficient stability and constancy for the evolution of a highly ordered state. The methodology described here attempts to meet most of the required conditions for development of highly ordered mixed cultures in which the anti-entropic force is perhaps at its highest. From an industrial point of view the system which produces this ordered state will be needed as the preferred method of strain maintenance and inoculum preparation. The system thereby retains its usefulness once production-scale fermentation is underway and continues to justify its initial capital cost and complexity.

The fermentation unit, like the industrial scale process it is intended to simulate, is of course subject to process upsets of various kinds, all of which affect the cultural conditions to some degree. Some cultures produce divergent oscillations in population density after relatively minor process upsets; others produce convergent oscillations leading to stability (Tsuchiya *et al.*, 1972). Clearly the latter cultures are to be preferred from an industrial point of view.

The stability of a culture can be tested systematically by introducing single step changes in any one of the measured variables in our fermentation system. Such a programme of work represents the early stages of process optimization, which, for complex mixed interactive cultures, can only be achieved on a well monitored and controlled system such as that described here. Full discussion of this point is beyond the scope of this chapter (see Larsen, 1976).

For use in a non-aseptic fermentation, a derived culture should be able to resist challenges from potential "contaminant" organisms such as those found in activated sewage (Johnson, 1972). In our system all non-methane-oxidizing species present in the fermenter are dependent on the activity of the methane oxidizers for their nutrition and survival, a feature which aids the development of a highly ordered contaminant-free culture.

Other uses of the system

Our system is ideal for isolation of gas-utilizing species including N_2-fixers, autotrophs (CO_2- fixing) and hydrogen users, and for gas-producing species (anaerobic methanogenic bacteria).

In some applications (e.g. simultaneous consumption of several carbon and energy sources) turbidostatic operation is preferable to chemostatic operation. In a turbidostatic mixed culture all cohabiting species grow in conditions of nutrient excess at their maximum rate, the ratios of individual species reflecting their individual growth rate maxima. The technical difficulties of continuous measurement of turbidity can be overcome in our system by linking the medium input pump to the rise and fall in any growth-related variable. For example, it is a simple matter to arrange that the medium pump operates at a high flow rate when a pre-set level of methane consumption or oxygen consumption or CO_2 accumulation has been reached. The induced wash-out of cells reduces, say, the methane consumption and the medium flow stops. Methane consumption rises again and the cycle is repeated. The selection pressures in such a system favour species with high growth rate maxima but not necessarily high yields on substrate. Turbidostat systems are normally wasteful of nutrients (which are lost in the effluent streams) but in a gas recycle system all gaseous nutrients are conserved. A chemostatic culture which for some reason has lost stability can sometimes be rescued by changing to turbidostat operation for a while until the cause of the upset has been located. This changeover requires simple electrical wiring alterations only and can be usefully accommodated on a multiple changeover switch.

Adaptability of control equipment

It is emphasized that the process controllers listed here represent a "basic kit" which can be assembled to meet several different applications. For example the system can be used as an oxystat in which the current output from a dissolved oxygen probe is put through a potentiometer and the resultant voltage is used by the Leeds and Northrup proportional electronic controller to make adjustments to the partial pressure of oxygen in the gas circulation loop in order to maintain a constant dissolved oxygen tension. Similarly, the pH meter-controller can be used with a probe for measurement of E_h in anaerobic fermentations. The meter's recorder output can be fed to the Leeds and Northrup controller in order to adjust the partial pressure of nitrogen in the circulating gas, thereby maintaining the desired E_h value. Alternatively, gas production—as occurs in the anaerobic digestion of organic matter—can be monitored by the system pressure controller in which the normal control action is reversed, i.e. pressure rise causes a pneumatic valve to open, thereby releasing gas for collection externally. The flow of gas can be continuously measured by the flow transmitter and also can be used as the control signal for turbidostat cultivation.

Any signal input which is not immediately suitable for use can usually be transduced into an appropriate electrical or pneumatic signal to match the controllers listed.

Application to production of secondary metabolites

The new role of continuous open systems for bypassing the normal cell regulation of secondary metabolite production and allowing over-production of such metabolites, is discussed by Demain (1972). Our fermentation system allows the isolation and maintenance of suitable cultures for this important development.

References

Ainsworth, G. (1973). Sludge Treatment—the current trends. *Proc. Biochem.*, (1), 11.

Bungay, H. R. (1968). Microbial interactions in continuous culture. *Adv. appl. Microbiol.*, **10**, 269.

Davies, D. S. (1974). Raw materials for chemical industry. *Chemtech*, March, 135.

Demain, A. L. (1972). Cellular and environmental factors affecting the synthesis and excretion of metabolites. *J. appl. Chem. Biotechnol.*, **22**, 345.

Dworkin, M. & Foster, J. W. (1956). Studies on *Pseudomonas methanica* (shöngen) nov. comb. *J. Bacteriol.*, **72**, 646.

GRAY, K. R., BIDDLESTONE, A. J. & SHERMAN, K. (1971). A review of composting. *Proc. Biochem.*, **6**, 32.

HARDER, W. (1975). The utilisation of mixed substrates by microorganisms. *6th Int. Symp. Study Group, Continuous Culture of Microorganisms.* Oxford, 20th–26th July.

HARRISON, D. E. F., WILKINSON, T., WREN, S. & HARWOOD, J. (1975). Mixed bacterial cultures as a basis for continuous production of SCP from C_1 compounds. *6th Int. Symp. Study Group, Continuous Culture of Micro organisms.* Oxford, 20th–26th July.

HOCKENHULL, J. (1975). The fermentation pilot plant and its aims. *Adv. appl. Microbiol.*, **19**, 187.

HUMPHREY, A. E. (1972). The role of fermentation technology in modern society. *Proc. IVth Int. Ferm. Symp.* (Terui, G., ed.), p. 13.

JANNASH, M. W. & MATELES, R. I. (1974). Experimental bacterial ecology studied in continuous culture. *Adv. microb. Physiol.*, **11**, 165.

JOHNSON, M. (1972). Techniques for selection and evaluation of cultures for biomass production. *Proc. IVth Int. Ferm. Symp.* (Terui, G., ed.), p. 473.

JOST, J. L., DRAKE, J. F., FREDERICKSON, A. G. & TSUCHIYA, H. M. (1973). Interactions of *Tetrahymena pyriformis*, *Escherichia coli*, *Azotobacter vinelandii*, and glucose in minimal medium. *J. Bacteriol.*, **113**, 834.

LARSEN, V. F. (1976). Ph.D. Thesis, Univ. of Strathclyde, Glasgow, Scotland.

LARSEN, V. F., SPIVEY, M. J. & HOLDOM, R. S. (1976). Automatic membrane-filtration system for the on-demand supply of large volumes of sterile medium in continuous culture. *Biotech. Bioeng.*, **18**, 1433.

LEADBETTER, E. R. & FOSTER, J. W. (1958). Studies on some methane utilising bacteria. *Arch. Mikrobiol.*, Nov., 1473.

LEHNINGER, A. L. (1965). *Bioenergetics.* New York: Benjamin.

MEERS, J. L. (1973). Growth of bacteria in mixed cultures. *Crit. Revs Microbiol.*, **2**, 139.

MILLER, T. L. (1966). Microbial utilisation of hydrocarbons. *Diss. Abstr.* (B), **27**, 1063.

PEITERSEN, N. (1975). Cellulose and protein production from mixed cultures of *Trichoderma viride* and a yeast. *Biotech. Bioeng.*, **17**, 1291.

PIRT, S. J. (1975). Principles of microbe and cell cultivation. Oxford: Blackwell, p. 75.

RICICIA, J. (1969). Multistage systems. In *Methods in microbiology*, Vol. 2 (Norris, J. B. & Ribbons, D., eds). London and New York: Academic Press, p. 329.

SHEEHAN, B. T. & JOHNSON, M. J. (1971). Production of bacterial cells from methane. *Appl. Microbiol.*, **21**, 511.

SNEDECOR, B. & COONEY, C. L. (1974). Thermophilic mixed culture of bacteria utilising methane for growth. *Appl. Microbiol.*, **27**, 112.

SPIVEY, M. J. (1973). Ph.D. Thesis, Univ. of Strathclyde, Glasgow, Scotland.

SUKATSCH, D. A. & JOHNSON, M. J. (1972). Bacterial cell production from hexadecane at high temperatures. *Appl. Microbiol.*, **23**, 543.

TSUCHIYA, H. M., JOST, J. L. & FREDRICKSON, A. G. (1972). Intermicrobial symbiosis. *Proc. IVth Int. Ferm. Symp.* (Terui, G., ed.), p. 43.

VAN DIJKEN, J. P. & HARDER, W. (1975). Growth yields of microorganisms on methanol and methane. A theoretical study. *Biotech. Bioeng.*, **17**, 15.

VARY, P. S. & JOHNSON, M. J. (1967). Cell yields of bacteria grown on methane. *Appl. Microbiol.*, **18**, 1473.

VELDKAMP, H. (1975). Mixed culture studies with the chemostat. *6th Int. Symp.*

Study Group. Continuous Culture of Microorganisms. Oxford, July 20th–26th.
WHITTENBURY, R., (1969). Microbial utilisation of methane. *Process Biochem.*, Jan., 51.
WILKINSON, T. G. & HARRISON, D. E. F. (1972). The affinity for methane and methanol of mixed cultures grown on methane in continuous culture. *J. Appl. Bacteriol.*, **36**, 309.
WILKINSON, T. G., TOPIWALE, H. H. & HAMER, G. (1974). Interactions in a mixed bacterial population growing on methane in continuous culture. *Biotech. Bioeng.*, **16**, 41.

Appendix 1

Recent uses of mixed cultures

Application	Reference
Fungus and yeast (one of two strains). Improved yield of biomass on alkali-treated barley straw. No adverse effect on food value of *Trichoderma* sp.	Peitersen (1975)
Two bacteria. Growth at 65° on hexadecane only if both species present.	Sukatsch and Johnson (1972)
Three bacteria, all required for maximum biomass yield at 50–56°. Growth up to 65°. Lower nucleic acid content, increased protein: RNA ratio.	Snedecor and Cooney (1974)
Single cell protein by non-aseptic fermentation—a review.	Jannasch and Mateles (1974)
Growth on methane with increased process stability, yield, growth rate, resistance to contamination. Foaming reduced.	Harrison *et al.* (1975)

Enrichments in a Chemostat

C. M. Brown

University of Dundee, Dundee, Scotland

D. C. Ellwood and J. R. Hunter

Microbiological Research Establishment
Porton, Wiltshire, England

Introduction

Most natural environments contain a large variety of microbes in numbers which reflect the habitat and the relative abilities of the individual organisms to compete for the nutrients available. In most investigations the total population is sparse and if the organisms of interest are present in low relative numbers then some enrichment is necessary before isolation is attempted. As enrichment usually means altering the environment in order to increase the population size then this process is highly selective. The constituent organisms of the enriched population will depend upon the chemical composition of the medium used together with the temperature, E_h, pH, presence of selective substrates or inhibitors, etc. and many species present initially will be lost due to their failure to compete in the enrichment system.

In the closed, batch culture used traditionally for enrichments the concentrations of nutrients are usually initially high to produce a large population of organisms. In many instances some succession of species occurs as the chemical environment changes with time. Those organisms able to grow fastest at any one time predominate and selection is governed by the maximum specific growth rates of the species involved.

In this paper we outline the use of the chemostat as an enrichment system and show how this method may be used to select for organisms unlikely ever to predominate in the batch system.

The Chemostat as an Enrichment System

The theory of the chemostat has been the subject of many articles and reviews (see for example Herbert *et al.*, 1956; Herbert, 1958; Tempest, 1970). This system exploits the fact that given a constant temperature, pH, etc., the specific growth rate of a microbial population depends upon the concentration of a growth-limiting nutrient in the culture medium. In its simplest form this relationship may be described by the Monod equation

$$\mu = \mu m \frac{S}{K_S + S} \tag{1}$$

where μ is the specific growth rate, S the concentration of limiting nutrient in the culture, μm the growth rate produced at saturating values of S and K_S a saturation constant being numerically equal to that concentration of S producing a μ of $\frac{1}{2}$ μm. K_S is a measure of the affinity of the organism for the nutrient involved and is in the order mg litre^{-1} for carbohydrates and μg litre^{-1} for amino acids (see Herbert, 1958). μ and S are therefore related by a saturation curve of the type shown in Figs 1 and 2.

In a chemostat one nutrient in the incoming medium is usually maintained at a relatively low concentration. S is fixed by the rate of addition of fresh medium (the dilution rate) and in turn controls μ at some point on the μ/S curve. In the chemostat then μ is maintained at values below μm, fixed by the culture dilution rate.

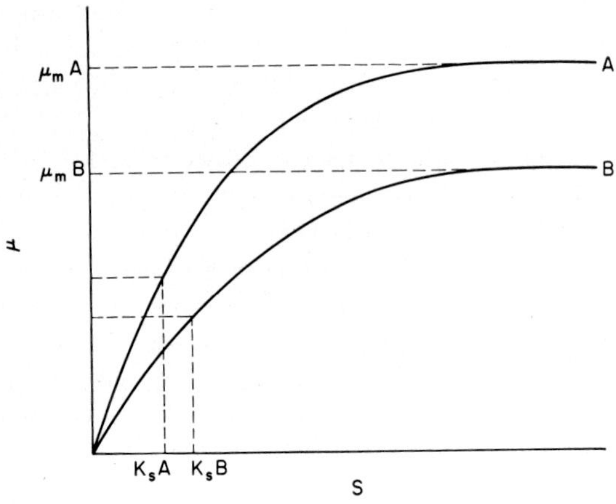

FIG. 1. The relationship between μ and S. Organism A will outgrow organism B at all growth rates, in both batch and continuous cultures.

The behaviour of mixed cultures in the chemostat has also been described (Powell, 1958; Veldkamp, 1970; Jannasch and Mateles, 1974). The system is highly selective and in a mixed population competing for a single limiting nutrient then the outcome, given constant temperature, pH, etc. and no interaction, will be determined by the μ/S relationships of the organisms involved. Figure 1 shows a system involving organism A with a higher μm and K_S than organism B. In this example A will

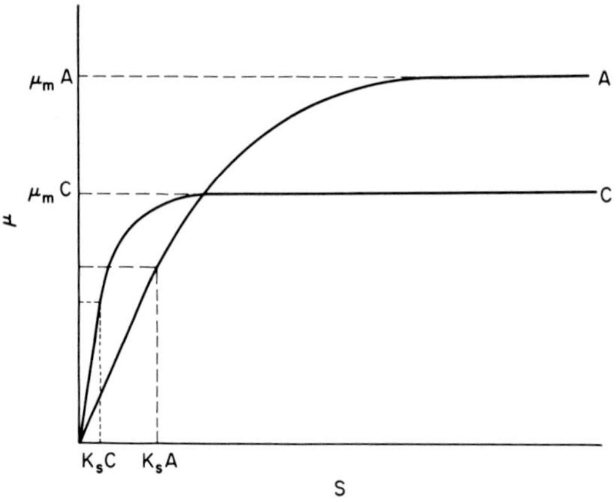

FIG. 2. The relationship between μ and S. Organism A will outgrow organism C at high growth rates. C will outgrow A only in continuous culture at low dilution rates.

outgrow B at any value of μ imposed by the dilution rate and will predominate both in batch and chemostat culture enrichments in the medium involved. If, however, the system represented by Fig. 2 is considered then with organisms A and C the μ/S curves cross and while A will outgrow C at high growth rates C will outgrow A at low growth rates. In a batch culture with saturating values of S selection is based solely on the μm and A always predominates. In a chemostat the outcome of competition depends upon the dilution rate imposed and by using low rates it is possible to enrich for the low K_S, low μm organisms of type C which never predominate in batch culture enrichments. Examples of μ/S curves which cross and therefore result in different organisms predominating at different growth rates may be found in the papers of Jannasch (1967), Harder and Veldkamp (1971), Meers (1971), Veldkamp and Jannasch (1972), Veldkamp and Kuenen (1973) and Jannasch and Mateles (1974).

Examples of Chemostat Enrichments

Jannasch pioneered the use of continuous culture enrichment methods and suggested that those systems had three particular advantages (Jannasch, 1965, 1967).

(1) No succession of species occurred and if there was no wall growth or interaction then the predominance of one species increased with time.

(2) The growth advantages of the successful competitor are not dependent on substrate specificity but on the particular growth parameters of the organism and the cultural conditions provided. If these parameters are known and stable then enrichment is reproducible.

(3) Enrichments may be carried out in the presence of extremely low concentrations of the limiting nutrient and therefore at low population densities.

These advantages are especially true for enrichment from aquatic environments and (1) and (3) are of general application. Substrate specificity (2 above) might be the property of interest in many investigations. For example in Novicks' original experiments his aim was to isolate spontaneous mutants with increased μm on tryptophan and to estimate mutation rates (Novick and Szilard, 1950). The ability to grow well in a carbon-limited environment on some novel hydrocarbon or other energy source, the ability to carry out a detoxification reaction, grow well in the presence of a growth inhibitor or produce a particular enzyme are other instances where selection on the basis of substrate specificity may be carried out usefully in a chemostat. In some cases, however, there may be little advantage over batch culture methods. To this growth rate selection system may be added other parameters such as variations in temperature, pO_2, etc. Harder and Veldkamp (1971) have shown how the temperature of incubation influences competition between different organisms with different temperature—μ/S characteristics.

Jannasch (1967) has described a number of enrichments from sea water and showed that within a range of dilution rates and limiting nutrient concentrations these systems showed a high degree of reproducibility. In these experiments the medium used was filter sterilized sea water supplemented with a carbon source (glucose, glycerol or lactate), phosphate (at pH 7·8) and NH_4Cl in a C:N:P ratio of 10:4:1 by weight. This medium produced a carbon-limited environment. Samples were removed at intervals of half a retention time and plated out on media containing ten times the nutrient concentrations of the liquid media. In some instances, especially at high nutrient concentrations (e.g. 10 or 100 mg $litre^{-1}$ lactate) and at low growth rates no species predominated, the cultures were visibly turbid and there was a significant amount of wall growth. This problem of wall growth is discussed below. In most instances, however, a single bacterial species accounted for 90% or so of the population and was isolated and characterized. Depending upon the dilution rate and nutrient concentration used the predominant species were *Vibrio*, *Spirillum*, *Micrococcus*, *Pseudomonas* or *Achromobacter*. To test the validity of the theory of this type of enrichment two of these isolates were used in a competition experiment. One, a *Spirillum* sp., was the predominant organism in an enrichment at a dilution rate of 0·1 h^{-1} and the other, a *Pseudomonas* sp. which predominated at a dilution rate of 0·2 h^{-1}. When made to compete as a mixed culture the *Spirillum* outgrew the *Pseudomonas* at low dilution rates and the *Pseudomonas* the *Spirillum* at high dilution rates. One problem encountered with these organisms was the apparent instability of their growth parameters when cultures were maintained at nutrient concentrations higher than present in the enrichment. This may be due to selection in batch culture of mutants with an increased μm and Jannasch (1967) recommended the use of liquid media of very low nutrient concentration for maintenance.

The experiments described above were carried out in carbon-limited cultures but the system is applicable to other limiting nutrients. For example Veldkamp and Kuenen (1973) quote experiments with a phosphate limited enrichment of ditch water carried out at two different dilution rates. At a dilution rate of 0·03 h^{-1} a *Spirillum* sp. outgrew all other bacteria while at 0·3 h^{-1} an unidentified rod predominated.

Isolation of Mixed Cultures from the Oral Cavity

One of the commonest of human diseases is dental caries, the familiar decay of teeth. Organisms occur in dental plaque, a film which covers the tooth surface. It was of interest to study the interactions of these plaque bacteria when growing at rates which might be relevant to the situation *in vivo*, i.e. mean generation times of 8–24 h.

A sample of dental plaque was inoculated into a complex medium to which 1% glucose had been added. Previous work had shown that when *Streptococcus mutans* strain Ingbritt was grown in this medium the culture was glucose limited (Ellwood *et al.*, 1974). The culture was grown overnight as a batch culture in a Porton-type chemostat and then put on flow at a dilution rate of 0·05 h^{-1}, i.e. a mean generation time of 14 h. When steady state was reached the volatile acids were analysed by gas liquid chromatography (g.l.c.). Substantial quantities of acetic and propionic acids were formed together with some isobutyric, valeric, isocaproic and caproic acids. Presumably it was these latter acids that gave the culture its characteristic revolting odour which was reminiscent of that found on opening a dental root abscess. Although these acids were produced the culture did not require the addition of alkali to maintain the pH at 6·5 because sufficient ammonia was produced by the culture itself. The acids produced were studied as a function of dilution rate and it was found that acetate and propionate were still produced but the amount of iso-acids present fell as the dilution rate increased (Ellwood *et al.*, 1972).

The situation was quite complex and much more work needs to be done. Interestingly it was also found that when samples of plaque from different mouths (adult and juvenile) were used as inocula, substantially the same results were obtained.

Isolation of Mixed Cultures from Marine Environments

Recently we carried out enrichments from sea water in an artificial sea water medium (Brown and Stanley, 1972) in which the nitrogen source was made the limiting nutrient (Whittaker and Brown, unpublished). These experiments were designed to enrich for organisms possessing uptake/reduction systems for nitrate of high substrate affinity as earlier batch enrichment isolates had shown relatively poor affinity (Brown *et al.*, 1975). Using an initial medium nitrogen concentration of 71 μg litre^{-1} the residual concentration was 1–5 μg litre^{-1}. Enrichments were performed in chemostats of 1 litre capacity of the type described by Baker (1968) and initially were filled with sea water. Artificial sea water was added continuously at a dilution rate of 0·02 to 0·03 h^{-1} and the temperature maintained at 10°. A population density of approximately 5×10^7 ml^{-1} was produced with little wall growth. After about five days small motile Gram-negative rod-shaped bacteria predominated. These were *Pseudomonas* sp. and one such isolate had a K_S for nitrate uptake of 5×10^{-5} M. Another chemostat was set up with the same sea water initially but in media containing 50 times the nutrient concentration of that described. The bacterial population was much more diverse than in the more dilute medium and no single species predominated. In addition this type of enrichment invariably also contained protozoa which appeared to graze on the bacteria. In one experiment ciliates were found, but in general the protozoal population consisted of small flagellates many of which were probably *Bodo* spp. Bacterial numbers of up to 5×10^8 ml^{-1} and protozoal numbers of up to 5×10^5 ml^{-1} were recorded. The data on the relative numbers of bacteria and protozoa suggest oscillations of the type reported by Curds (1970). The bacteria tended to form clumps making counting difficult and in addition there was extensive growth on the walls of the culture vessel. Obviously wall growth makes the whole system heterogeneous and the selection pressures outlined above will be exerted on the "suspended" rather than the "surface" population although the two must inevitably be related. The surface population was much more diverse in terms of different bacterial types than the population in suspension.

There is some literature on the ability of marine bacteria to stick to surfaces (including glass) by means of an extracellular polysaccharide "adhesive" and the role of this material in the colonization of these bacteria has been considered by Zobell (1943), Corpe (1970a, b), Marshall *et al.* (1971a, b), Fletcher and Floodgate (1973). A nitrogen-limited environment of the type used above appears to be ideal growth conditions for the production of such a polysaccharide (in that N-limitation is known to promote the formation of intra- and extracellular polysaccharides) and such material was visible in the form of extensive capsular substance. Glass slides suspended within the growth vessel quickly became covered with bacteria, aided presumably by this capsular material. In aquatic environments the ability to grow on surfaces may be an ecological advantage and the use of a glass slide or some other surface within a chemostat enrichment system appears to offer a useful system for isolation of organisms possessing this characteristic. As previously mentioned growth as a film is of importance in the attachment of bacteria to the tooth surface. It is thought that the organisms in dental plaque secrete enzymes which are able to convert sucrose to poly-glucan with which the organisms form a matrix giving rise to the bacterial film. Destruction of the teeth results from the action of acid secreted by the organisms giving rise to dental caries.

Growth at a Surface

In many natural systems there appears to be some constraint on the growth of bacteria because the bacterial population is low. It seems reasonable to assume that the number of bacteria present is related to the capacity of the environment to provide nutrients for growth. If all nutrients are supplied in excess then bacteria would grow at a rate which was optimal for the other conditions imposed by the environment, e.g. temperature and pH, etc. Growth rates in nature may be low even in optimal environmental conditions and it seems likely that nutrient supply could be the rate limiting process for bacterial growth. The physico-chemical laws of absorption suggest that the greatest concentration of nutrients would be found at the interfaces between liquid and solids in any sysetm. Hence it would follow that bacteria would grow better in this area of increased concentration of nutrient and a film would build up. The growth of the film as a whole will be limited by the diffusion of substrates into the film.

The maximum thickness of a film of *Escherichia coli*, not oxygen limited, with one side exposed to air at 101 kPa was estimated to be approximately 40 μm by Pirt (1967). He also estimated the thickness of fungal growth limited by glucose concentration of 10 g litre^{-1} to gave an active layer of about 1000 μm. If the film is thicker than the maximum thickness of an active film then the layer below this may be considered to be an area of inactivity or a different limiting substrate is now controlling that activity. The intermediate layers of the film could be oxygen limited, and anaerobic growth with carbon limitation could apply at the bottom of the film. So that in general there could be a series of zones in a film in which growth was controlled by a number of different limiting nutrients. These would depend on the diffusion gradient and of course on the uptake of a limiting nutrient by organisms growing at different parts of the film. This growth of films in a fermenter has been recently reviewed by Atkinson and Fowler (1974).

Discussion

We have been primarily concerned in this paper with the question of enrichment methods using the chemostat technique. We have pointed out that systems in which the competing species do not interact are reasonably well understood and these ideas have already had important implications in our understanding of microbial ecology. The situation with respect to mixed cultures which do interact is as yet not well understood, and much further work is required. We have also discussed the question of microbial growth in films which appears to be widespread in nature as exemplified by the very different areas of oral and marine microbiology. The study of the mechanisms of film formation seems to be very important but as yet is little understood.

A number of problems arise in the use of chemostats to isolate mixed cultures of interacting species. The usual practice is to inoculate a fermenter containing its working volume of medium and allow the organisms to grow as a batch culture before putting the chemostat on flow.

Thus the reservations we have expressed about the use of batch culture techniques apply to this situation. In the subsequent operation of the chemostat only those organisms which have survived the batch culture will grow and a false picture of the original inoculum would emerge. This situation could arise due to the phenomenon of substrate accelerated death (S.A.D.) (Postgate and Hunter, 1964), if the inoculum, as seems likely when taken from a natural environment, was growing in a nutrient-limiting situation. This could be partially overcome by the addition of cyclic AMP to the original batch culture. Perhaps a better method would be to put the inoculum into the fermenter first, suspended in a non-nutrient-containing solution, and then add the medium at a slow rate to this suspension. This may be used to provide a spectrum of cultural conditions according to the volume of suspending menstruum relative to the working volume of the reactor. In all cases, the nutrients at first will be in such a low concentration that S.A.D. is not likely to occur. When the volume in the fermenter reached overflow then the chemostat operation would start, and slow growing organisms would have had a chance to survive. It might be necessary to re-inoculate the chemostat several times to ensure that the organisms which should predominate under the conditions imposed by the chemostat would do so.

The cultures isolated by this technique would be a reflection of the nutrient limitation and as such may not bear much of a relationship to the natural system from which they were derived. However, if nutrient limitation was important in that natural system it should be possible for this to be identified. Once this was achieved then it should be possible to get enriched cultures which would allow us to mimic the natural system under study.

References

Atkinson, B. & Fowler, H. W. (1974). The significance of microbial film in fermenters. *Adv. biochem. Engin.*, **3**, 221.

Baker, K. (1968). Low cost continuous culture apparatus. *Lab. Practice*, **17**, 817.

Brown, C. M. & Stanley, S. O. (1972). Environment-mediated changes in the cellular content of the 'pool' constituents and their associated changes in cell physiology. *J. appl. Chem. Biotechnol.*, **22**, 363.

Brown, C. M., Macdonald-Brown, D. S. & Stanley, S. O. (1975). Inorganic nitrogen metabolism in marine bacteria: Nitrate uptake and reduction in a marine pseudomonad. *Mar. Biol.*, **31**, 7.

Corpe, W. A. (1970a). An acid polysaccharide produced by a primary film-forming marine bacterium. *Develop. indust. Microbiol.*, **11**, 402.

Corpe, W. A. (1970b). Attachment of marine bacteria to solid surfaces. In *Adhesion in biological systems* (Manly, R. S., ed.). New York and London: Academic Press.

Curds, C. R. (1970). A continuous culture. A method for the determination of food consumption by ciliated protozoa. In *Proceedings of the symposium on methods of study of soil ecology* (Philipson, J., ed.). Paris: UNESCO, p. 127.

Ellwood, D. C., Longyear, V. M. C. & Hunter, J. R. (1972). Growth of mixed cultures of organisms derived from human dental plaque in a chemostat. *J. Gen. Microbiol,* **71**, 10.

Ellwood, D. C., Hunter, J. R. & Longyear, V. M. C. (1974). Growth of *Streptococcus mutans* in a chemostat. *Arch. of Oral Biol.*, **19**, 659.

Fletcher, M. & Floodgate, G. D. (1973). An electron-microscope demonstration of an acidic polysaccharide involved in the adhesion of a marine bacterium to solid surfaces. *J. gen. Microbiol.*, **74**, 325.

Harder, W. & Veldkamp, H. (1971). Competition of marine psychrophilic bacteria at low temperatures. *Antonie van Leeuwenhoek*, **37**, 51.

Herbert, D. (1958). Some principles of continuous culture. In *Recent progress in microbiology*. 7th International Congress for Microbiology, Stockholm. Oxford: Blackwell, p. 381.

Herbert, D., Elsworth, R. & Telling, R. C. (1956). The continuous culture of bacteria: a theoretical and experimental study. *J. gen. Microbiol.*, **14**, 601.

Jannasch, H. W. (1965). Continuous culture in microbial ecology. *Lab. Practice*, **14**, 1162.

Jannasch, H. W. (1967). Enrichments of aquatic bacteria in continuous culture. *Archiv fur Mikrobiol.*, **59**, 165.

Jannasch, H. W. & Mateles, R. I. (1974). Experimental bacterial ecology studied in continuous culture. *Adv. microb. Physiol.*, **11**, 165.

Marshall, K. C., Stout, R. & Mitchell, R. (1971a). Mechanism of the initial events in the sorption of marine bacteria to surfaces. *J. gen. Microbiol.*, **68**, 337.

Marshall, K. C., Stout, R. & Mitchell, R. (1971b). Selective sorption of bacteria from sea water. *Can. J. Microbiol.*, **17**, 1413.

Meers, J. L. (1971). Effect of dilution rate on the outcome of chemostat mixed culture experiments. *J. gen. Microbiol.*, **67**, 359.

Novick, A. & Szilard, L. (1950). Experiments with the chemostat on spontaneous mutation of bacteria. *Proc. Nat. Acad. Sci.* (*Washington*), **36**, 708.

Pirt, S. J. (1967). A kinetic study of the mode of growth of surface colonies of bacteria and fungi. *J. gen. Microbiol.*, **47**, 181.

Postgate, J. R. & Hunter, J. R. (1964). Accelerated death of *Aerobacter aerogenes* starved in the presence of growth-limiting substrates, *J. gen. Microbiol.*, **34**, 459.

Powell, E. O. (1958). Criteria for the growth of contaminants and mutants in continuous culture. *J. gen. Microbiol.*, **18**, 259.

Tempest, D. W. (1970). The continuous cultivation of microorganisms. 1. Theory of the chemostat. In *Methods in Microbiology*, Vol. 2 (Norris, J. R. & Ribbons, D. W., eds). London and New York: Academic Press, p. 259.

Veldkamp, H. (1970). Enrichment cultures of prokaryotic organisms. *Methods in Microbiology*, Vol. 3A (Norris, J. R. & Ribbons, D. W., eds). London and New York: Academic Press, p. 305.

Veldkamp, H. & Jannasch, H. W. (1972). Mixed culture studies with the chemostat. *J. appl. Chem. Biotechnol.*, **22**, 105.

Veldkamp, H. & Kuenen, J. G. (1973). The chemostat as a model system for ecological studies. *Bull. ecol. Res. Commission.* (*Stockholm*), **17**, 247.

ZoBell, C. E. (1943). The effect of solid surfaces upon bacterial activity. *J. Bacteriol.*, **46**, 39.

Subject Index